SpringerBriefs in Electrical and Computer Engineering

Signal Processing

Series editors

Woon-Seng Gan, Singapore, Singapore
C.-C. Jay Kuo, Los Angeles, USA
Thomas Fang Zheng, Beijing, China
Mauro Barni, Siena, Italy

More information about this series at http://www.springer.com/series/11560

Yushu Zhang · Yong Xiang
Leo Yu Zhang

Secure Compressive Sensing in Multimedia Data, Cloud Computing and IoT

Yushu Zhang
School of Information Technology
Deakin University
Burwood, Melbourne, VIC, Australia

Leo Yu Zhang
School of Information Technology
Deakin University
Burwood, Melbourne, VIC, Australia

Yong Xiang
School of Information Technology
Deakin University
Burwood, Melbourne, VIC, Australia

ISSN 2191-8112   ISSN 2191-8120   (electronic)
SpringerBriefs in Electrical and Computer Engineering
ISSN 2196-4076   ISSN 2196-4084   (electronic)
SpringerBriefs in Signal Processing
ISBN 978-981-13-2522-9   ISBN 978-981-13-2523-6   (eBook)
https://doi.org/10.1007/978-981-13-2523-6

Library of Congress Control Number: 2018954023

This Springer imprint is published by the registered company Springer Nature Singapore Pte Ltd.
The registered company address is: 152 Beach Road, #21-01/04 Gateway East, Singapore 189721, Singapore

To my beloved Sue and Aite
—Yushu Zhang

To my beloved Shan, Angie, and Daniel
—Yong Xiang

To my parents and my beloved family
—Leo Yu Zhang

Preface

Compressive sensing is always being a hot topic, since only requiring fewer samples can achieve the same recovery effect for the signal compared with the traditional Shannon–Nyquist sampling theorem. It has increasing applications in various fields such as image processing, pattern recognition, cloud computing, and Internet of things. There is a very promising application prospect that compressive sensing can be regarded as a cryptosystem when the measurement matrix acts as a key, thus achieving simultaneous sampling, compression, and encryption. Exploiting compressive sensing for security aspects has been widely studied in both theory and application. In particular, the increasing works about multimedia data security, cloud computing security, and Internet of things security based on compressive sensing have been investigated in recent years. This book gives a comprehensive and systematic introduction for secure compressive sensing with applications in multimedia data, cloud computing, and Internet of things, which will help readers well grasp the knowledge of this field, understand the corresponding development trend, and explore emerging research opportunities, challenges, and applications.

Melbourne, Australia

Yushu Zhang
Yong Xiang
Leo Yu Zhang

Acknowledgements

The work is supported by the National Natural Science Foundation of China (NSFC) (No. 61702221).

Contents

Chapter 1
Compressive Sensing

Abstract Since compressive sensing (CS) theory has come into the world, it has been widely applied in many fields. It was claimed that both sampling and compression can be performed simultaneously to reduce the sampling rate at the expense of a high computation complexity at the reconstruction stage. By virtue of the sparsity, a signal, which is randomly projected at the encoder side, can be reconstructed by searching the optimal solution of an under determined linear system at the decoder side. In information security field, the CS can be utilized for multimedia data security, cloud computing security, internet of things (IoT) security, etc.

1.1 Theory of Compressive Sensing

Compressive sensing has received extensive research attention in the last decade [1–4]. By utilizing the fact that natural signals are either sparse or compressible, the CS theory demonstrates that such signals can be faithfully recovered from a small set of linear, nonadaptive measurements, allowing sampling at a rate lower than that required by the Nyquist-Shannon sampling theorem.

Suppose an N-dimensional signal $\mathbf{x} \in \mathbb{R}^N$ is expressed as

$$\mathbf{x} = \sum_{i=1}^{N} x_i \boldsymbol{\psi}_i = \boldsymbol{\Psi}\theta, \tag{1.1}$$

which means that $\mathbf{x}$ could be sparsely represented in a certain domain by the transform matrix $\boldsymbol{\Psi} := [\psi_1, \ \psi_2, \ \ldots, \ \psi_N]$ with each column vector $\psi_i \in \mathbb{R}^N, i = 1, 2, \ldots, N$. We can say that $\mathbf{x}$ is exactly k-sparse if there are at most k non-zero coefficients in the $\boldsymbol{\Psi}$ domain. Instead of sampling $\mathbf{x}$ directly, we take a small number of CS measurements. Let $\Phi := [\varphi_1, \ \varphi_2, \ \cdots, \ \varphi_N]$ denote an $M \times N$ matrix with $M \ll N$. Then M non-adaptive linear samples $\mathbf{y}$ can be obtained by

$$\mathbf{y} = \Phi\mathbf{x}. \tag{1.2}$$

© The Author(s), under exclusive license to Springer Nature Singapore Pte Ltd. 2019
Y. Zhang et al., *Secure Compressive Sensing in Multimedia Data,*
Cloud Computing and IoT, SpringerBriefs in Signal Processing,
https://doi.org/10.1007/978-981-13-2523-6_1

The resultant CS measurements **y** are used for the recovery of the original signal by solving the following convex optimization problem

$$\min \|\theta\|_1 \; s.t. \; \mathbf{y} = \mathbf{\Phi\Psi}\theta$$

$$(or \; in \; noisy \; situation: \; \|\mathbf{\Phi\Psi}\theta - \mathbf{y}\|_2 \le \varepsilon) \tag{1.3}$$

to obtain $\tilde{\mathbf{x}} = \mathbf{\Psi}\tilde{\theta}$, where $\tilde{\theta}$ is the solution of the optimization problem and $\tilde{\mathbf{x}}$ represents the reconstructed signal.

One of the central problems in CS framework is the selection of a proper measurement matrix $\mathbf{\Phi}$ satisfying the Restricted Isometry Property (RIP).

Definition 1 ([5]) Matrix $\mathbf{\Phi}$ satisfies the RIP of order k if there exists a constant $\delta_k \in [0, 1]$ such that

$$(1 - \delta_s) \|\mathbf{x}\|_2^2 \le \|\mathbf{\Phi x}\|_2^2 \le (1 + \delta_s) \|\mathbf{x}\|_2^2 \tag{1.4}$$

for all k-sparse signals **x**.

Candès and Tao [6] proposed that a matrix following the Gaussian or Bernoulli distribution satisfies RIP with a overwhelming probability at the sparsity $k \le \mathcal{O}(M/\log N)$. The randomly selected Fourier basis also retains RIP with a overwhelming probability at the sparsity $k \le \mathcal{O}(M/(\log N)^6)$.

1.2 Applications of Compressive Sensing

So far, there are numerous works on applications of CS and there will be more and more works in future. This book focuses on information security applications of CS in terms of multimedia data security, cloud computing security, and IoT security. As a typical form of multimedia data, image is encrypted using CS frequently combined with other cryptographic techniques, such as chaos theory and optical transform, to enjoy higher security level and realize simultaneous encryption and compression [7–28]. For example, Zhou et al. designed a novel key-controlled measurement matrix, which is established by leveraging the circulant matrices and manipulating the original row vectors of the circulant matrices with Logistic chaos map [7]. They also adopted the partial Hadamard matrix as measurement matrix conducted by chaotic map and the generated measurements are further scrambled [8]. Later, Zhou et al. suggested an efficient image compression-encryption method based on hyper-chaotic system and two-dimensional (2D) CS to reduce the possible transmission burden [29]. The authors in [12] developed a simultaneous image encryption and compression scheme based on random convolution and random subsampling. This scheme can be tailored for a single image unlike the existing joint optical encryption and compression schemes for multiple images and has similar architecture with double random phase encoding (DRPE). Lang and Zhang came up with a unique perspective of transmitting only a few measurements intermittently chosen from the masks rather than the real keys and the tremendous masks codes [10], since the parameters of the chaotic maps can be inferred from the received measurements without error so that the correct random phase-amplitude masks can be obtained and used for decrypting the encoded information. The DRPE based block CS was designed for image encryption, which aims to encrypt each image block using a chaos-based random phase encoding in fractional Fourier domain [11]. In addition, from the viewpoint that digital devices can only store the samples at a finite precision, Zhang et al. suggested a joint quantization and diffusion approach for the real-valued measurements based on the distribution of measurements of natural images sensed by structurally random ensemble [13]. The issue of creating a CS-based symmetric cipher under the key reuse circumstance was tackled in [14]. Specifically, a bi-level protected CS mode was projected by taking use of the advantage of the non-RIP measurement matrix construction. It was validated that the mode can be resistant to common attacks even a fixed measurement matrix is used multiple times. In [16], Zhang et al. proposed some possible encryption models for CS and then demonstrated random permutation is an acceptable permutation with overwhelming probability, which can effectively relax the RIP for parallel CS (PCS). Random permutation is used for creating a secure PCS scheme and the corresponding security analysis indicates the asymptotic spherical secrecy. In order to resist chosen-plaintext attacks, Fay introduced the counter mode of operation to CS-based encryption and it can achieve probabilistic encryption [17]. Besides CS-based image encryption, there exist some research works on image watermarking [30–36], image hiding [37–41], image hashing [42–45], image authentication [46–49] and others [50–53] based on CS. For video data, CS-based privacy protection and watermarking frameworks were proposed in [54–57]. On the audio side, CS based hash algorithm was designed in [58].

An emerging technology for providing multimedia services and applications is multimedia cloud computing [59]. Under the background of cloud multimedia computing, the CS technique can offer privacy-preserving multimedia cloud computing [60], outsourcing of image reconstruction service [61], multimedia data storage and sharing [62–64], healthcare monitoring system [65], and parallel outsourcing of sparse reconstruction service [66, 67]. In cloud security scenario, privacy-assured multimedia cloud computing based on CS and sparse representation was investigated in [60], which discussed some compressive multimedia applications, including multimedia compression, adaptation, editing/manipulation, enhancement, retrieval, and recognition. Wang et al. [61] proposed a privacy-preserving outsourcing of image reconstruction service from CS in cloud [61]. Different domain technologies were synthesized to fulfill the prospective on the aspects of security, efficiency and complexity. They further widened the outsourcing of image construction service to healthcare monitoring system [65]. In order to simultaneously perform secure watermark detection and privacy-protected multimedia data storage in a cloud computing application scenario, Wang et al. designed such a framework based on CS and secure multiparty computation protocols under the hypothesis of the semi-honest adversary model [62]. The hybrid cloud was used for providing secure big image data storage and share service for users [63], in which the core idea is to partition each image into a small set of sensitive data and a large set of insensitive data, which are securely stored in the private cloud and the public cloud, respectively. In order to guarantee data privacy and meanwhile maintain image management, a novel outsourced image reconstruction and identity authentication service was outsourced to the public cloud [64], in which both the techniques of signal processing in the CS domain and computation outsourcing are adopted. Outsourcing sparse reconstruction service and sparse robustness decoding service to multi-clouds in parallel was described in [66, 67] by the assumption that multi-clouds cannot collude with each other in private. The privacy of the original signal can be guaranteed, since each cloud only has a small amount of information of both the measurements and asymmetric support-set.

With the development of IoT, the existing security and privacy problems in it are receiving more and more attentions [68–70]. The CS has some applications in IoT security [71, 72]. The adaptive CS can provide lightweight compression and encryption for resource-limited smart objects, which are the basic blocks in IoT system, by utilizing the information of some smart objects and adapting the CS measurement conditions for the remaining smart objects [73]. Besides of CS, the frequency was taken into consideration for the physical layer security [74], since the information is possibly leaked in the static environment. This scheme has a high efficiency and a strong security level because of the utilizations of circulant matrix as the measurement matrix and a binary resilient function. To address the issue of the static application scenario, the frequency-selective feature of the wireless channel was used in [75] to enlarge the entropy of the measured channel and accelerate the rate of generating keys from physical layer. The multimedia IoT [76] is becoming increasingly popular with the era of multimedia data and social networks, which involves two core problems, low-complexity sampling and confidentiality protection. To solve these two problems, a framework was presented in [77] based on chaotic CS. The encryption scheme supports batch image processing, suitable for multimedia big data. In addition, the encryption mechanism has a two-layer architecture including chaotic measurement matrix and permutation-diffusion, thus preserving the confidentiality well. The IoT data need to be stored in cloud server, which will give rise to the problem of privacy explosion due to the untrust of cloud. Based on CS, a secure interaction scheme between IoT and the cloud was proposed in [78]. Random compressed encryption is used to capture the raw data and uploads them to the cloud server. Such encryption mode belongs to a kind of multiplicative coefficient perturbation such that some statistical values are directly calculated over encrypted domain without decryption. This kind of encryption also supports secure data insertion and accurate raw decryption in cloud-enabled IoT scenarios. In addition, there exist some works focusing on compressive detection with secrecy guarantees [79, 80], i.e., to directly detect the compressive measurements without reconstruction while maintaining the secrecy.

1.3 Organization

The book is arranged as follows. In Chap. 2, we address the issue of secure CS. Firstly, how CS can be regarded as a cryptosystem is stated. Then, some relevant works are overviewed from the design and analysis of CS cryptosystem, respectively. In Chap. 3, multimedia data security based on CS is investigated, which involves the combination of CS and multimedia encryption techniques, PCS for multi-dimensional multimedia data, the involvement of image processing techniques, and double protection mechanism for CS Sampling. Cloud computing security related to CS is discussed in Chap. 4, in which two CS reconstruction schemes including privacy-preserving sparse reconstruction service and secure sparse robustness decoding service are designed. Chapter 5 is about CS-based IoT security, where a secure low-cost CS sampling framework and a secure data storage and sharing framework are described. Finally, concluding remarks and future works are provided in Chap. 6.

References

1. D.L. Donoho, Compressed sensing. IEEE Trans. Inf. Theory **52**(4), 1289–1306 (2006)
2. R.G. Baraniuk, Compressive sensing. IEEE Signal Process. Mag. **24**(4), 118–121 (2007)
3. E.J. Candès, M.B. Wakin, An introduction to compressive sampling. IEEE Signal Process. Mag. **25**(2), 21–30 (2008)
4. E.J. Candès, J. Romberg, T. Tao, Robust uncertainty principles: exact signal reconstruction from highly incomplete frequency information. IEEE Trans. Inf. Theory **52**(2), 489–509 (2006)
5. E.J. Candès, T. Tao, Decoding by linear programming. IEEE Trans. Inf. Theory **51**(12), 4203–4215 (2005)
6. E.J. Candès, T. Tao, Near-optimal signal recovery from random projections: Universal encoding strategies? IEEE Trans. Inf. Theory **52**(12), 5406–5425 (2006)
7. N. Zhou, A. Zhang, F. Zheng, L. Gong, Novel image compression-encryption hybrid algorithm based on key-controlled measurement matrix in compressive sensing. Opt. Laser Techn. **62**, 152–160 (2014)
8. N. Zhou, A. Zhang, J. Wu, D. Pei, Y. Yang, Novel hybrid image compression-encryption algorithm based on compressive sensing. Optik **125**(18), 5075–5080 (2014)
9. S.N. George, D.P. Pattathil, A novel approach for secure compressive sensing of images using multiple chaotic maps. J. Opt. **43**(1), 1–17 (2014)
10. J. Lang, J. Zhang, Optical image cryptosystem using chaotic phase-amplitude masks encoding and least-data-driven decryption by compressive sensing. Opt. Commun. **338**, 45–53 (2015)
11. H. Liu, D. Xiao, Y. Liu, Y. Zhang, Securely compressive sensing using double random phase encoding. Optik **126**(20), 2663–2670 (2015)
12. Y. Zhang, L.Y. Zhang, Exploiting random convolution and random subsampling for image encryption and compression. Electron. Lett. **51**(20), 1572–1574 (2015)
13. L.Y. Zhang, K.-W. Wong, Y. Zhang, Q. Lin, Joint quantization and diffusion for compressed sensing measurements of natural images, in *Proceedings of IEEE International Symposium on Circuits and Systems, ISCAS* (2015), pp. 2744–2747
14. L.Y. Zhang, K.-W. Wong, Y. Zhang, J. Zhou, Bi-level protected compressive sampling. IEEE Trans. Multimed. **18**(9), 1720–1732 (2016)
15. J. Li, J.S. Li, Y.Y. Pan, R. Li, Compressive optical image encryption. Sci. Rep. **5**, 10374 (2015)

16. Y. Zhang, J. Zhou, F. Chen, L.Y. Zhang, K.-W. Wong, H. Xing, D. Xiao, Embedding crypto-graphic features in compressive sensing. Neurocomput. **205**, 472–480 (2016)
17. R. Fay, Introducing the counter mode of operation to compressed sensing based encryption. Inf. Process. Lett. **116**(4), 279–283 (2016)
18. Y. Zhang, J. Zhou, F. Chen, L.Y. Zhang, D. Xiao, B. Chen, L. Xiaofeng, A block compressive sensing based scalable encryption framework for protecting significant image regions. Int. J. Bifurcat. Chaos **26**(11), 1650191 (2016)
19. H. Huang, X. He, Y. Xiang, W. Wen, Y. Zhang, A compression-diffusion-permutation strategy for securing image. Signal Process. **150**, 183–190 (2018)
20. D. Zhang, X. Liao, B. Yang, Y. Zhang, A fast and efficient approach to color-image encryption based on compressive sensing and fractional Fourier transform. Multimed. Tools Appl. **77**(2), 2191–2208 (2018)
21. J. Chen, Y. Zhang, L.Y. Zhang, On the security of optical ciphers under the architecture of compressed sensing combining with double random phase encoding. IEEE Photonics J. **9**(4), 1–11 (2017)
22. X. Chai, Z. Gan, Y. Chen, Y. Zhang, A visually secure image encryption scheme based on compressive sensing. Signal Process. **134**, 35–51 (2017)
23. N. Zhou, J. Yang, C. Tan, S. Pan, Z. Zhou, Double-image encryption scheme combining DWT-based compressive sensing with discrete fractional random transform. Opt. Commun. **354**, 112–121 (2015)
24. X. Chai, X. Zheng, Z. Gan, D. Han, Y. Chen, An image encryption algorithm based on chaotic system and compressive sensing. Signal Process. **148**, 124–144 (2018)
25. Y. Zhang, L.Y. Zhang, J. Zhou, L. Liu, F. Chen, X. He, A review of compressive sensing in information security field. IEEE Access **4**, 2507–2519 (2016)
26. X. Li, X. Meng, X. Yang, Y. Yin, Y. Wang, X. Peng, W. He, G. Dong, H. Chen, Multiple-image encryption based on compressive ghost imaging and coordinate sampling. IEEE Photonics J. **8**(4), 1–11 (2016)
27. G. Hu, D. Xiao, Y. Wang, T. Xiang, An image coding scheme using parallel compressive sensing for simultaneous compression-encryption applications. J. Visual Commun. Image Represent. **44**, 116–127 (2017)
28. G. Hu, D. Xiao, Y. Wang, T. Xiang, Q. Zhou, Securing image information using double random phase encoding and parallel compressive sensing with updated sampling processes. Opt. Lasers Eng. **98**, 123–133 (2017)
29. N. Zhou, S. Pan, S. Cheng, Z. Zhou, Image compression-encryption scheme based on hyper-chaotic system and 2D compressive sensing. Opt. Laser Tech. **82**, 121–133 (2016)
30. G. Valenzise, M. Tagliasacchi, S. Tubaro, G. Cancelli, M. Barni, A compressive-sensing based watermarking scheme for sparse image tampering identification, in *Proceedings of 16th IEEE International Conference on Image Processing, ICIP* (2009), pp. 1265–1268
31. X. Zhang, Z. Qian, Y. Ren, G. Feng, Watermarking with flexible self-recovery quality based on compressive sensing and compositive reconstruction. IEEE Trans. Inf. Forensics Sec. **6**(4), 1223–1232 (2011)
32. H.-C. Huang, F.-C. Chang, C.-H. Wu, W.-H. Lai, Watermarking for compressive sampling applications, in *Proceedings of Eighth International Conference on Intelligent Information Hiding and Multimedia Signal Processing, IIH-MSP* (2012), pp. 223–226
33. I. Orovic, A. Draganic, S. Stankovic, Compressive sensing as a watermarking attack, in *Proceedings of 21st Telecommunications Forum, TELFOR* (2013), pp. 741–744
34. I. Orovic, S. Stankovic, Combined compressive sampling and image watermarking, in *Proceedings of 55th International Symposium on ELMAR* (IEEE, 2013), pp. 41–44
35. D. Xiao, M. Deng, Y. Zhang, Robust and separable watermarking algorithm in encrypted image based on compressive sensing. J. Electron. Inf. Techn. **37**(5), 1248–1254 (2015)
36. H. Liu, D. Xiao, R. Zhang, Y. Zhang, S. Bai, Robust and hierarchical watermarking of encrypted images based on compressive sensing. Signal Process.-Image Commun. **45**, 41–51 (2016)
37. W. Li, J.-S. Pan, L. Yan, C.-S. Yang, H.-C. Huang, Data hiding based on subsampling and compressive sensing, in *Proceedings of Ninth International Conference on Intelligent Information Hiding and Multimedia Signal Processing* (2013), pp. 611–614

38. J.-S. Pan, W. Li, C.-S. Yang, L.-J. Yan, Image steganography based on subsampling and compressive sensing. Multimed. Tools Appl., 1–15 (2014)
39. D. Xiao, S. Chen, Separable data hiding in encrypted image based on compressive sensing. Electron. Lett. **50**(8), 598–600 (2014)
40. G. Hua, Y. Xiang, G. Bi, When compressive sensing meets data hiding. IEEE Signal Process. Lett. **23**(4), 473–477 (2016)
41. M. Li, D. Xiao, Y. Zhang, Reversible data hiding in block compressed sensing images. ETRI J. **38**(1), 159–163 (2016)
42. L.-W. Kang, C.-S. Lu, C.-Y. Hsu, Compressive sensing-based image hashing, in *Proceedings of 16th IEEE International Conference on Image Processing, ICIP* (2009), pp. 1285–1288
43. M. Tagliasacchi, G. Valenzise, S. Tubaro, Hash-based identification of sparse image tampering. IEEE Trans. Image Process. **18**(11), 2491–2504 (2009)
44. R. Sun, W. Zeng, Secure and robust image hashing via compressive sensing. Multimed. Tools Appl. **70**(3), 1651–1665 (2014)
45. H. Liu, D. Xiao, Y. Xiao, Y. Zhang, Robust image hashing with tampering recovery capability via low-rank and sparse representation. Multimed. Tools Appl. **75**(13), 7681–7696 (2016)
46. H. Suzuki, M. Suzuki, T. Urabe, T. Obi, M. Yamaguchi, N. Ohyama, Secure biometric image sensor and authentication scheme based on compressed sensing. Appl. Opt. **52**(33), 8161–8168 (2013)
47. H. Suzuki, M. Takeda, T. Obi, M. Yamaguchi, N. Ohyama, K. Nakano, Encrypted sensing for enhancing security of biometric authentication, in *Proceedings of 13th Workshop on Information Optics (WIO)* (2014), pp. 1–3
48. D. Xiao, M. Deng, X. Zhu, A reversible image authentication scheme based on compressive sensing. Multimed. Tools Appl. **74**(18), 7729–7752 (2015)
49. J. Chen, Z.-L. Zhu, C. Fu, L.-B. Zhang, Y. Zhang, Information authentication using sparse representation of double random phase encoding in fractional fourier transform domain. Optik **136**, 1–7 (2017)
50. L.-W. Kang, C.-Y. Lin, H.-W. Chen, C.-M. Yu, C.-S. Lu, C.-Y. Hsu, S.-C. Pei, Secure transcoding for compressive multimedia sensing, in *Proceedings of 18th IEEE International Conference on Image Processing, ICIP* (2011), pp. 917–920
51. J.K. Pillai, V.M. Patel, R. Chellappa, N.K. Ratha, Secure and robust iris recognition using random projections and sparse representations. IEEE Trans. Pattern Anal. Mach. Intell. **33**(9), 1877–1893 (2011)
52. L. Liu, A. Wang, C.-C. Chang, Z. Li, A novel real-time and progressive secret image sharing with flexible shadows based on compressive sensing. Signal Process.-Image Commun. **29**(1), 128–134 (2014)
53. J. Qi, X. Hu, Y. Ma, Y. Sun, A hybrid security and compressive sensing-based sensor data gathering scheme. IEEE Access **3**, 718–724 (2015)
54. M. Cossalter, M. Tagliasacchi, G. Valenzise, Privacy-enabled object tracking in video sequences using compressive sensing, in *Proceedings of Sixth IEEE International Conference on Advanced Video and Signals-based Surveillance* (2009), pp. 436–441
55. L. Tong, F. Dai, Y. Zhang, J. Li, D. Zhang, Compressive sensing based video scrambling for privacy protection, in *Proceedings of IEEE Visual Communications and Image Processing, VCIP* (2011), pp. 1–4
56. X. Chen, H. Zhao, A novel video content authentication algorithm combined semi-fragile watermarking with compressive sensing, in *Proceedings of Second International Conference on Intelligent System Design and Engineering Application, ISDEA* (2012), pp. 134–137
57. L.G. Jyothish, V. Veena, K. Soman, A cryptographic approach to video watermarking based on compressive sensing, arnold transform, sum of absolute deviation and svd, in *Proceedings of Annual International Conference on Emerging Research Areas* (2013), pp. 1–5
58. G. Valenzise, G. Prandi, M. Tagliasacchi, A. Sarti, Identification of sparse audio tampering using distributed source coding and compressive sensing techniques. J. Image Video Process. **2009**, 1 (2009)

59. W. Zhu, C. Luo, J. Wang, S. Li, Multimedia cloud computing. IEEE Signal Process. Mag. **28**(3), 59–69 (2011)
60. L.-W. Kang, K. Muchtar, J.-D. Wei, C.-Y. Lin, D.-Y. Chen, C.-H. Yeh, privacy-preserving multimedia cloud computing via compressive sensing and sparse representation, in *Proceedings of International Conference on Information Security and Intelligent Control, ISIC* (2012), pp. 246–249
61. C. Wang, B. Zhang, K. Ren, J. Wang, Privacy-assured outsourcing of image reconstruction service in cloud. IEEE Trans. Emerg. Top. Comput. **1**(1), 166–177 (2013)
62. Q. Wang, W. Zeng, J. Tian, A compressive sensing based secure watermark detection and privacy preserving storage framework. IEEE Trans. Image Process. **23**(3), 1317–1328 (2014)
63. Y. Zhang, H. Huang, Y. Xiang, L.Y. Zhang, X. He, Harnessing the hybrid cloud for secure big image data service. IEEE Internet Things J. **4**(5), 1380–1388 (2017)
64. G. Hu, D. Xiao, T. Xiang, S. Bai, Y. Zhang, A compressive sensing based privacy preserving outsourcing of image storage and identity authentication service in cloud. Inf. Sci. **387**, 132–145 (2017)
65. C. Wang, B. Zhang, K. Ren, J.M. Roveda, C.W. Chen, Z. Xu, A privacy-aware cloud-assisted healthcare monitoring system via compressive sensing, in *Proceedings of INFOCOM* (2014), pp. 2130–2138
66. Y. Zhang, J. Zhou, L.Y. Zhang, F. Chen, X. Lei, Support-set-assured parallel outsourcing of sparse reconstruction service for compressive sensing in multi-clouds, in *Proceedings of International Symposium on Security and Privacy in Social Networks and Big Data, SocialSec* (2015), pp. 1–6
67. Y. Zhang, J. Zhou, Y. Xiang, L.Y. Zhang, F. Chen, S. Pang, X. Liao, Computation outsourcing meets lossy channel: secure sparse robustness decoding service in multi-clouds. IEEE Trans. Big Data (in press, 2017)
68. J. Granjal, E. Monteiro, J.S. Silva, Security for the internet of things: a survey of existing protocols and open research issues. IEEE Commun. Surveys Tuts. **17**(3), 1294–1312 (2015)
69. J. Lin, W. Yu, N. Zhang, X. Yang, H. Zhang, W. Zhao, A survey on internet of things: architecture, enabling technologies, security and privacy, and applications. IEEE Internet Things J. **4**(5), 1125–1142 (2017)
70. Y. Yang, L. Wu, G. Yin, L. Li, H. Zhao, A survey on security and privacy issues in internet-of-things. IEEE Internet Things J. **4**(5), 1250–1258 (2017)
71. A. Mukherjee, Physical-layer security in the internet of things: sensing and communication confidentiality under resource constraints. Proc. IEEE **103**(10), 1747–1761 (2015)
72. A. Fragkiadakis, E. Tragos, A. Makrogiannakis, S. Papadakis, P. Charalampidis, M. Surligas, Signal processing techniques for energy efficiency, security, and reliability in the IoT domain, in *Internet of Things (IoT) in 5G Mobile Technologies* (Springer, 2016), pp. 419–447
73. A. Fragkiadakis, P. Charalampidis, E. Tragos, Adaptive compressive sensing for energy efficient smart objects in IoT applications, in *4th International Conference on Wireless Communications, Vehicular Technology, Information Theory and Aerospace and Electronic Systems, VITAE* (IEEE, 2014), pp. 1–5
74. N. Wang, T. Jiang, W. Li, S. Lv, Physical-layer security in internet of things based on compressed sensing and frequency selection. IET Commun. **11**(9), 1431–1437 (2017)
75. M. Wilhelm, I. Martinovic, J.B. Schmitt, Secure key generation in sensor networks based on frequency-selective channels. IEEE J. Sel. Areas Commun. **31**(9), 1779–1790 (2013)
76. S.A. Alvi, B. Afzal, G.A. Shah, L. Atzori, W. Mahmood, Internet of multimedia things: vision and challenges. Ad Hoc Netw. **33**, 87–111 (2015)
77. Y. Zhang, Q. He, Y. Xiang, L.Y. Zhang, B. Liu, J. Chen, Y. Xie, Low-cost and confidentiality-preserving data acquisition for internet of multimedia things. IEEE Internet Things J. (in press, 2017)
78. W. Xue, C. Luo, G. Lan, R.K. Rana, W. Hu, A. Seneviratne, Kryptein: a compressive-sensing-based encryption scheme for the internet of things, in *Proceedings of 16th ACM/IEEE International Conference on Information Processing in Sensor Networks, IPSN* (2017), pp. 169–180

79. B. Kailkhura, S. Liu, T. Wimalajeewa, P.K. Varshney, Measurement matrix design for com-
    pressed detection with secrecy guarantees. IEEE Wireless Commun. Lett. **5**(4), 420–423 (2016)
80. B. Kailkhura, T. Wimalajeewa, P.K. Varshney, Collaborative compressive detection with phys-
    ical layer secrecy constraints. IEEE Trans. Signal Process. **65**(4), 1013–1025 (2017)

Chapter 2
Secure Compressive Sensing

Abstract This chapters introduces how CS becomes a cryptosystem and then overviews the designs and analyses of some CS cryptosystems.

2.1 Compressive Sensing as a Cryptosystem

The reason of CS as a cryptosystem is to enable the measurement matrix to be a key known by the encoder and the decoder. This idea is originally mentioned in a foundational work of CS [1], in which Candes and Tao suggested that the measurement vector obtained from random subspace linear projection can be treated as ciphertext since the unauthorized user would not be able to decode it unless he knows in which random subspace the coefficients are expressed. In this way, the entire CS scheme can be considered as a variant of symmetric cipher, where the signal to be sampled, the measurement vector and the measurement matrix are treated as the plaintext, the ciphertext and the secret key, respectively.

For better understanding, a contrast between CS and symmetric-key cipher is visually illustrated in Fig. 2.1, in which $\mathbf{f}$ represents a compressible signal. If the signal itself is sparse, it will be changed into a simple case, i.e., just delete $\mathbf{\Psi}$ and $\mathbf{\Psi}^T$. The basic model of CS is shown in the upper half of Fig. 2.1 and there are two major aspects: measurements taking and signal recovery. From the perspective of symmetric-key cipher, measurements taking involves an encryption algorithm and signal recovery is associated with a decryption algorithm.

2.2 Design of Compressive Sensing Cryptosystem

The use of CS for security purposes was first outlined in one of the foundation papers [1]. It was also mentioned in [2] that a pseudorandom basis can provide a weak form of secrecy and in [3] that there is a relationship between the encryption matrix and one-time pad cipher. Rachlin and Baron formally investigated the security of CS as a

Y. Zhang et al., *Secure Compressive Sensing in Multimedia Data,*
Cloud Computing and IoT, SpringerBriefs in Signal Processing,
https://doi.org/10.1007/978-981-13-2523-6_2

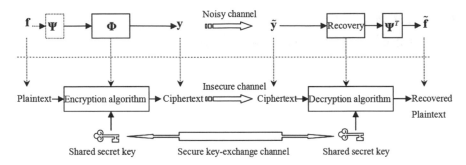

Fig. 2.1 A contrast between CS and symmetric-key cipher

cryptosystem with the measurement matrix as a key and found that CS cannot achieve Shannon's perfect secrecy but can guarantee the computational secrecy [4]. Brute force and structured attacks were found to be unsuccessful in breaking this cryptosystem [5], which also indicated that the robustness against additive noise can be provided. The security of Guassian measurement matrix was analysed by inspecting the mutual information between the plaintext and the ciphertext without the assumption of specific statistical distribution of plaintext [6]. The mutual information can evaluate the secrecy level of the CS cryptosystem to some extent and quantitatively reflect the leaked information from a viewpoint of ciphertext-only attack. Moreover, this cryptosystem can achieve a perfect secrecy defined by Bianchi et al. by means of a Gaussian i.i.d. matrix as measurement matrix and a normalization of measurements [7, 8]. Furthermore, circulant matrix implemented by fast Fourier transform was also introduced in [9] for the purpose of cost savings while maintaining the similar reconstruction performance as the Gaussian matrix. Differing a general security analysis that assumes that the key is absolutely secure, an interesting security analysis claimed that the key can be partly revealed for sparse plaintext by developing the prior sparsity knowledge of the plaintext [10]. The mutual information between the plaintext and the ciphertext was extended to that of the ciphertext, the key, and the plaintext. Combining the extended mutual information and the plaintext sparsity feature, the new perfect secrecy criteria including Shannon-sense and Wyner-sense are defined to measure the secrecy. Besides of these measurement matrices as the keys, some novel measurement matrices have been designed. For example, Cambareri et al. [11] designed a two-class information concealing system based on perturbing the measurement matrix, in which the first-class users can reconstruct the signal to its full resolution while the second-class ones are able to retrieve only a degraded version of the same signal. This two-class case was further extended to the multiclass case in [12].

2.3 Analysis of Compressive Sensing Cryptosystem

Some works [13–18] focus on the security analysis of the existing CS encryption schemes. Mangia et al. performed a security evaluation for rakenness-based CS, where the measurement matrix is not independent and identically distributed and this information is also known to the adversary [13]. The energy of the ciphertext is altered and the corresponding statistics also carry the information other than the energy of the plaintext. Result shows that it satisfies the asymptotic circular secrecy. Nevertheless, these security analyses aim at the viewpoint of the ciphertext-only attack, and the authors in [13, 15] performed a known-plaintext attack to the rakenness-based CS, which has a stronger analysis capacity compared with the ciphertext-only attack, since it can deal with some known plaintext/ciphertext pairs. A quantitative result was finally given. The perturbation-type measurement matrix [12] was taken into account by launching a known-plaintext attack [14], which shown that the candidate solutions matching a plaintext-ciphertext pair have an enormous number and therefore the attack can be well resisted, since it is not feasible to find out a true measurement matrix from a vast set. Attackers in the chosen-plaintext attack work in a stronger assumption than that of the known-plaintext attack. To have an resistance of the chosen-plaintext attack, Fay introduced the counter mode [16], parallelizability and self-synchronization [17] in CS, so that a set of keys was used for encrypting signals many times to attain the probabilistic encryption, no longer confined to one-time key scenario. In addition, an authenticated encryption mode was presented to cope with the attack of data integrity [18]. A message authentication code is first incorporated in the encryption process of CS and then the receiver re-computes this code, compares the extracted code and finally judges the behaviour of data integrity.

References

1. E.J. Candès, T. Tao, Near-optimal signal recovery from random projections: Universal encoding strategies? IEEE Trans. Inf. Theory **52**(12), 5406–5425 (2006)
2. M.F. Duarte, S. Sarvotham, M.B. Wakin, D. Baron, R.G. Baraniuk, Joint sparsity models for distributed compressed sensing, in *Proceedings of Workshop Signal Processing with Adaptive Sparse Structural Representations* (IEEE, 2005)
3. I. Drori, Compressed video sensing, in *Proceedings of BMVA Symposium on 3D Video-Analysis Display Applications* (2008)
4. Y. Rachlin, D. Baron, The secrecy of compressed sensing measurements, in *Proceedings of 46th Annual Allerton Conference on Communication, Control, and Computing*, Urbana-Champaign, IL (2008), pp. 813–817
5. A. Orsdemir, H.O. Altun, G. Sharma, M.F. Bocko, On the security and robustness of encryption via compressed sensing, in *Proceedings of IEEE Military Communications Conference (MILCOM)*, San Diego, CA (2008), pp. 1–7
6. S.A. Hossein, A. Tabatabaei, N. Zivic, Security analysis of the joint encryption and compressed sensing, in *Proceedings of 20th Telecommunications Forum (TELFOR)*, Belgrade (2012), pp. 799–802

7. T. Bianchi, V. Bioglio, E. Magli, On the security of random linear measurements, in *Proceedings of IEEE International Conference on Acoustics, Speech, and Signal Processing (ICASSP)*, Florence (2014), pp. 3992–3996

8. T. Bianchi, V. Bioglio, E. Magli, Analysis of one-time random projections for privacy preserving compressed sensing. IEEE Trans. Inf. Forensics Secur. **11**(2), 313–327 (2016)

9. T. Bianchi, E. Magli, Analysis of the security of compressed sensing with circulant matrices, in *IEEE International Workshop on Information Forensics and Security, WIFS* (IEEE, 2014), pp. 173–178

10. Z. Yang, W. Yan, Y. Xiang, On the security of compressed sensing-based signal cryptosystem. IEEE Trans. Emerg. Topics Comput. **3**(3), 363–371 (2015)

11. V. Cambareri, J. Haboba, F. Pareschi, R. Rovatti, G. Setti, K.-W. Wong, A two-class information concealing system based on compressed sensing, in *Proceedings of IEEE International Symposium Circuits and System, ISCAS* (2013), pp. 1356–1359

12. V. Cambareri, M. Mangia, F. Pareschi, R. Rovatti, G. Setti, Low-complexity multiclass encryption by compressed sensing. IEEE Trans. Signal Process. **63**(9), 2183–2195 (2015)

13. M. Mangia, F. Pareschi, R. Rovatti, G. Setti, Security analysis of rakeness-based compressed sensing, in *IEEE International Symposium on Circuits and Systems, ISCAS* (IEEE, 2016), pp. 241–244

14. V. Cambareri, M. Mangia, F. Pareschi, R. Rovatti, G. Setti, On known-plaintext attacks to a compressed sensing-based encryption: a quantitative analysis. IEEE Trans. Inf. Forensics Sec. **10**(10), 2182–2195 (2015)

15. M. Mangia, F. Pareschi, R. Rovatti, G. Setti, Low-cost security of IoT sensor nodes with rakeness-based compressed sensing: statistical and known-plaintext attacks. IEEE Trans. Inf. Forensics Security **13**(2), 327–340 (2018)

16. R. Fay, Introducing the counter mode of operation to compressed sensing based encryption. Inf. Process. Lett. **116**(4), 279–283 (2016)

17. R. Fay, C. Ruland, Compressive sensing encryption modes and their security, in *11th International Conference for Internet Technology and Secured Transactions, ICITST* (IEEE, 2016), pp. 119–126

18. R. Fay, C. Ruland, Compressed sampling and authenticated-encryption, in *Proceedings of 11th International ITG Conference on Systems, Communications and Coding, SCC* (VDE, 2017), pp. 1–6

Chapter 3
Multimedia Data Security

Abstract Multimedia data have high redundancy. The CS itself can provide encryption protection when acquiring multimedia signals. Meanwhile, a special advantage of CS is that it can reduce the redundancy. Thus, CS is a new popular encryption technique used for multimedia data encryption. This chapter aims to design secure and efficient multimedia data encryption algorithms based on CS. Firstly, combining chaos theory and optical transform, which are two common tools of being used for encryption, we summarize six CS-based multimedia data encryption frameworks. Then, we designed PCS for multi-dimensional multimedia data encryption to avoid the problem of the measurement matrix size expansion. Furthermore, image processing techniques are involved to construct a scalable encryption framework, which can better protect important image regions. Lastly, to resist plaintext attacks, a double protection mechanism for CS is put forward.

3.1 Combination of Compressive Sensing and Multimedia Encryption Techniques

In the area of image encryption, it is well known that chaos theory and optical transform are the two most widespread and important technologies [1–19], since chaotic systems possess some instinctive properties such as ergodicity, pseudo-randomness and sensitivity to initial conditions and control parameters and optical transforms are noted for their high speed, parallel processing and large storage memories. Till now, the characteristics of CS, dimensional reduction and random projection, have been utilized and integrated into image ciphers based on chaos or optics, which can achieve simultaneous compression and encryption of an image or multiple images [20–41]. In the following, some design frameworks and their corresponding analyses are investigated with respect to image ciphers based on CS. Specifically, our investigation proceeds from three aspects, image ciphers based on chaos and CS, image ciphers based on optics and CS, and image ciphers based on chaos, optics and CS. A total of six frameworks are put forward. Meanwhile, their analyses in terms of security, advantages, disadvantages, future research topics, etc. are given.

Y. Zhang et al., *Secure Compressive Sensing in Multimedia Data,*
Cloud Computing and IoT, SpringerBriefs in Signal Processing,
https://doi.org/10.1007/978-981-13-2523-6_3

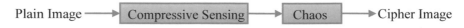

Plain Image ⟶ Compressive Sensing ⟶ Chaos ⟶ Cipher Image

Fig. 3.1 The sketch of Framework 1

3.1.1 Combination of Compressive Sensing and Chaos

Chaos and CS used in image ciphers are instantiated in two aspects: precedence relationship and nesting relationship. The former means one after the other while the latter is to embed one in the other. Thus, we have the following two basic frameworks.

Framework 1: Chaos-based encryption model is executed after CS, as shown in Fig. 3.1.

An image is firstly encrypted by CS, when the measurement matrix is considered as a secret key [42]. The acquired measurements are then encrypted by chaos-based models including permutation and diffusion. It should be noted that it is almost impossible to exploit chaos followed by CS due to the fact that chaos-based encryption always breaks the correlations between pixels and removes the redundancy farthest such that the sparsity which CS relies on cannot be guaranteed. In the following, we will take a case study for further illustration.

Case 1: Reference [43].

Huang et al. designed a parallel image encryption method based on CS and chaotic encryption models including Arnold scrambling, mixing, S-box, and chaotic lattice XOR. The original image is block-wise measured in parallel using Gaussian measurement matrix and quantized through the Lloyd quantizer. Then the data are reallocated for the purpose of the collision-free property that a communication unit exchanges data among the multiple processors without collision. Arnold scrambling is then used to permute the quantized measurements' positions followed by the mixing operation which makes a single alteration affect the final output. S-box substitution and block-wise XOR are finally used for diffusion.

Analysis: With respect to the only CS-based encryption scheme, i.e., measurement matrix as a key, the security has been discussed in [44–47], which demonstrated that the perfect secrecy is unachievable but the case of sensing signal with constant energy using Gaussian random matrix is perfectly secure. Moreover, it is not secure against chosen plaintext attack by choosing some particular sparse vector containing only one non-zero entry. However, in Framework 1, a great number of chaotic encryption techniques are able to remedy the security defect. However, there is no tight relationship between chaotic cipher and CS model reported so far. For the measurement values by CS, the chaotic cipher model designed should be tailored, for example, one can take account of the distribution of the measurements into the architecture of chaotic cipher. Otherwise, simple combination of CS and any chaotic encryption model can be adopted. Besides, there is a bunch of chaotic encryption models ensuring the security but a large proportion of them suffered by the problem

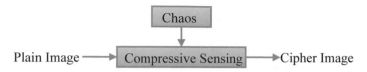

Fig. 3.2 The sketch of Framework 2

of high computation complexity and unfriendly hardware realization. Consequently, it is desirable to embed chaos in CS in some way.

Framework 2: Chaos is embedded in CS, as shown in Fig. 3.2.

In this model, the generation process of the measurement matrix is under the control of chaotic sequence, which is produced by chaotic system with initial values. Specifically, initial conditions or control parameters, which can be treated as secret keys, of the employed chaos system are used to generate the measurement matrix. This facilitates transmission and sharing that only requires a few values instead of the whole measurement matrix. At last, chaos-based CS samples images. There are many ways of implementing chaos to construct the measurement matrix, as illustrated in the following cases.

Case 1: Reference [22].

George and Pattathil employed eight 1D chaotic maps to generate the random measurement matrix. It has four stages: randomly selecting pairwise chaotic maps; calculating the initial values for the selected pair; iterating the selected chaotic maps; generating the random measurement matrix with zero mean and variance $1/N$. This matrix is adopted for block CS of images.

Case 2: Reference [48].

George and pattathil utilized linear feedback shift register (LFSR) for secure measurement matrix generation. Different states of LFSR are selected and normalized as the random entries of the measurement matrix. In order to withstand known plaintext attack, each block of an image is sampled by different measurement matrices, which are constructed by the LFSR system with a modulo division circuit and Logistic map. It avoided the memory overload.

Case 3: Reference [21].

In the case of the measurement matrix as a key, to make the key be easily distributed, stored and memorized, Zhou et al. created the improved circulant matrices by manipulating their original row vectors under the control of Logistic map. The original image is partitioned into four blocks and two of them in the adjacent locations exchange their pixels randomly. The random matrices used for exchanging the random pixel are tied up with measurement matrices.

Analysis: Framework 2 together with these three cases need to be aware of some problems. Prior to the image sampling, whether or not the newly generated measurement matrix by chaos satisfies the RIP condition should be demonstrated. The work [49], in which CS with chaotic sequence has been verified in theory, can serve as a

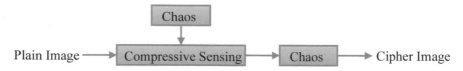

Fig. 3.3 The sketch of Framework 3

reference. Furthermore, the optimal reconstruction algorithm should be developed for better matching with the image sampling method. Meanwhile, the quantization is also worth considering for the purpose of real-time transmission. For security consideration, the initial values of the employed chaotic system require frequently alteration, otherwise they will be suffered by known-plaintext attack. Alternatively, to enhance the security, some appropriate chaotic encryption models may be added.

Framework 3: A hybrid of Framework 1 and Framework 2, as shown in Fig. 3.3.

Not only can chaos control CS, but also provide some encryption models after the CS sampling. As illustrated in Fig. 3.3, permutation and diffusion operations can be appropriately introduced to resist plaintext attacks after the CS sampling.

Case 1: Reference [20].

Zhou et al. adopted the partial Hadamard matrix controlled by chaos map as the measurement matrix. The image sampling is followed by a scrambling using the chaotic index sequence generated by the Logistic map.

Case 2: Reference [31].

Zhang et al. [31] suggested a scalable encryption framework based on block CS together with Sobel edge detector and cascade chaotic map for the purpose of protecting significant image regions. After an image is performed by Sobel edge detector, chaos-based structurally random matrix is applied to significant block encryption whereas chaos-based random convolution and subsampling are used for the remaining insignificant ones. This framework adopts lightweight subsampling and severe sensitivity encryption for the significant blocks and severe subsampling and lightweight robustness encryption for the insignificant ones in parallel.

Analysis: It is generally known that CS-based image ciphers are robust against noise, called robust encryption. However, in the above three basic frameworks, robust encryption may be broken by chaos effect. It is mainly due to the introduced chaotic encryption models. If only an permutation operation is employed, it does not affect the robustness; on the contrary, it may make the robustness more strong. For example, an acceptable permutation can relax the RIP condition [50], which may even reduce the block effects to some extent. If an diffusion operation is employed, the robustness must be broken since diffusion offers certain level of avalanche effect. In summary, Framework 2 is robust while Framework 1 and Framework 3 are uncertain. Apparently, a hybrid of Framework 1 and Framework 2 is more secure than their individual counterparts. As mentioned earlier, stronger security is at the expense of higher computation complexity and more chaotic encryption techniques, thereby a trade-off between security and overall complexity has to be examined.

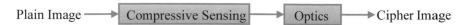

Fig. 3.4 The sketch of Framework 4

3.1.2 Combination of Compressive Sensing and Optics

The basic framework of optical image encryption is DRPE [51]. The most possible encryption combination of DRPE and CS is CS followed by DRPE.

Framework 4: CS is followed by DRPE, as shown in Fig. 3.4.

An image is firstly sampled by CS, and the acquired measurements are encrypted by the structure of DRPE. This can be validated by the following three cases.

Case 1: Reference [52].

Lu et al. proposed such a simple combination. The CS is used to directly encrypt and compress an image and then the DRPE is used for re-encryption.

Case 2: Reference [53].

Different from [52], Rawat et al. implemented an image quality enhancement procedure using iterative kernel steering regression algorithm [54] before performing Framework 4 and utilized the FrFT instead of Fourier transform for DRPE.

Case 3: Reference [55].

On the basis of Framework 4, Deepan et al. introduced space multiplexing [56] for multiple-image encryption. The multiple images are measured by CS respectively, and are then integrated by space multiplexing. It was claimed that this scheme is able to overcome the vulnerability [57–60] of classical DRPE due to its nonlinearity.

Analysis: Although Framework 4 realizes optical image compression and encryption, CS and DRPE only have a precedence relationship. It is of great significance to investigate their nesting relationship from the imaging prospective. A heuristic investigation is the work [61], which demonstrated the possibility of achieving super-resolution with a single exposure by combining DRPE and CS. Additionally, a bottleneck of CS in optics is to carry out the reconstruction algorithm using optical techniques, although it is easy for the sampling. As a consequence, the computing device has to be relied on. Besides of DRPE, some classic optical encryption techniques can also be infused in Framework 4 to be a novel scheme like multiple-image encryption [55]. This joint optical multiple-image compression and encryption work based on CS differs from the general ones [62–64]. In the end, another point of thought, some cryptographic features may be embedded in the existing CS-based imaging schemes [65–68].

3.1.3   Combination of Compressive Sensing, Chaos and Optics

When all the three techniques including chaos, optics and CS are applied, the corresponding framework is naturally a fusion of the above four frameworks.

Framework 5: A hybrid of Framework 1 and Framework 4, as shown in Fig. 3.5.

After the CS, one of chaotic encryption models and DRPE is exploited, and then the other follows.

Case 1: Reference [69].

Liu et al. employed a common chaotic permutation way, Arnold transformation, which is used to scramble the measurements obtained by CS. The results are again encrypted by DRPE, where two random phase masks are generated by sequences of irrational number.

Analysis: Generally speaking, this type of framework is secure enough, since three different encryption techniques can be fully utilized. However, there is nothing special but a simple combination. This approach will inevitably lead to a higher computation complexity. It is hopeful that they are further integrated.

Framework 6: A hybrid of Framework 2 and Framework 4, as shown in Fig. 3.6.

This framework means that chaos is embedded in not only CS but also optics. It is already clear that chaotic systems are able to supply the values to the orders of fractional transforms or random phase masks [70].

Case 1: Reference [71].

Liu et al. implemented chaos-based DRPE to encrypt CS measurements. In DRPE, the common Fourier transform is replaced by the FrFT. The measurement matrix and the random phase masks are furnished by pseudo-random sequences produced

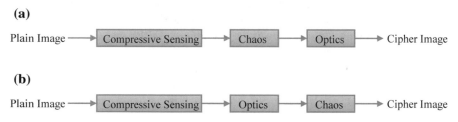

(a)

Plain Image ⟶ Compressive Sensing ⟶ Chaos ⟶ Optics ⟶ Cipher Image

(b)

Plain Image ⟶ Compressive Sensing ⟶ Optics ⟶ Chaos ⟶ Cipher Image

Fig. 3.5   The sketch of Framework 5

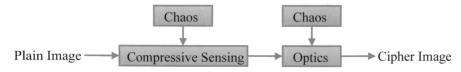

Fig. 3.6   The sketch of Framework 6

from Logistic map. In fact, this scheme can be further improved in the way that the fractional orders are also offered by Logistic map such that the whole encryption process is controlled by chaos.

Analysis: In this framework, chaos is embedded in both CS and optics, and there is a large advantage that both the huge measurement matrix and double random phase matrices are generated from a few initial values, which facilitates the sharing of the keys. Meanwhile, the introduction of chaos does not break the robustness of CS and DRPE architecture. Thus, Framework 6 is a robust encryption scheme but Framework 5 may not. More importantly, the initial value sensitiveness can be guaranteed. So far, Framework 6 is the preferred candidate for simultaneous compression and encryption of images.

In fact, there also exist a framework of image ciphers based on CS itself, as shown in [26, 27, 29, 30], which combine neither chaos nor optics. Finally, what is worth mentioning is a particular image cryptosystem using chaotic phase-amplitude masks and least-data-driven decryption by CS [23]. The highlight lies in that it transmits only a few measurements intermittently chosen from the masks rather than the real keys and the enormous mask codes. In the decryption side, utilizing the theories in [72, 73] and refining the series expansions, the decoder can infer the correct parameters of the chaotic map. This scheme motivates us to fully excavate some new proposed theories on the relationships among CS, chaos and optics for further study image cipher design.

3.2 Parallel Compressive Sensing for Multi-dimensional Multimedia Data

3.2.1 Random Permutation Meets Parallel Compressive Sensing

Traditionally, a multidimensional signal needs to be reshaped into an 1D signal prior to sampling using CS. Nevertheless, such a transformation makes the required size of the sensing matrix dramatically large and increases the storage and computational complexity significantly. To solve this problem, Fang et al. proposed a novel solution of PCS [50], which reshapes the multidimensional signal into a 2D signal and samples the latter column by column with the same sensing matrix. Moreover, a so-called acceptable permutation can effectively relax the RIP for PCS.

Definition 1 ([50]) For a 2D sparse signal $\mathbf{X}$ with sparsity vector $\mathbf{s} = [s_1, s_2, \ldots, s_N]$ satisfying $\|\mathbf{s}\|_1 = s$, where s_j is the sparsity level of the j-th column of $\mathbf{X}$, a permutation $\mathbf{P}(\bullet)$ is called acceptable for $\mathbf{X}$ if the Chebyshev norm of the sparsity vector of $\mathbf{P}(\mathbf{X})$ is smaller than $\|\mathbf{s}\|_\infty$ of $\mathbf{X}$.

When a 2D s-sparse signal is exactly reconstructed by using PCS, a sufficient condition is given by the following lemma.

Lemma 1 ([50]) *Consider a 2D* **s**-*sparse signal* **X**, *if the RIP of order* $\|\mathbf{s}\|_\infty$ *holds for the sensing matrix* $\mathbf{\Phi}$, *i.e.,* $\delta_{2\|\mathbf{s}\|_\infty} < \sqrt{2} - 1$, *then* **X** *can be exactly reconstructed using PCS scheme.*

This lemma implies that with respect to PCS, the RIP requirement of the sensing matrix $\mathbf{\Phi}$ at a given reconstruction quality is related to $\|\mathbf{s}\|_\infty$. A zigzag-scan per-mutation is considered acceptable in relaxing the RIP condition before using the PCS [50], but it is tailored for the sparse signal following a layer model. We gen-eralize the permutation for the 2D sparse signal whose distribution is unknown in advance. Assume that $\mathbf{P}(\bullet)$ is a random permutation operation, then $\mathbf{X}^* = \mathbf{P}(\mathbf{X})$, where $\mathbf{X}^* \in \mathbb{R}^{M \times N}$ is a permuted 2D signal with sparsity vector $\mathbf{s}^*$. Observing the relationship between random and acceptable permutations, we have the following theorem.

Theorem 1 *For a 2D sparse signal* **X**, *if the distribution of the sparsity level in each column is not sufficiently uniform (*$\|\mathbf{s}\|_\infty = \sigma \cdot \lceil \frac{s}{N} \rceil$*, where σ is assumed to be not less than 2.72 but $\|\mathbf{s}\|_\infty \ll M$), then the random permutation $\mathbf{P}(\bullet)$ can be an acceptable permutation with overwhelming probability.*

Proof If $\|\mathbf{s}^*\|_\infty \leq \|\mathbf{s}\|_\infty$, i.e., $\Pr\{\mathbf{P}(\bullet) \ is \ acceptable\} = 0$, meaning that each col-umn of **X** tends to have similar sparsity levels, $\mathbf{P}(\bullet)$ does not work. However, such an **X** has relaxed the RIP requirement for PCS without permutation. Thus, we consider the **X** where the distribution of the sparsity level in each column is not sufficiently uniform, which, more importantly, accords with the feature of a natural signal. Each element in **X** will be randomly located at any index of $\mathbf{X}^*$, that is, the transition of all the indices from **X** to $\mathbf{X}^*$ yields the uniform distribution. Each non-zero element of **X** appears in each column of $\mathbf{X}^*$ with equal probability $\frac{1}{N}$. This has a strong resem-blance to the classical probability problem of s balls and N boxes. The probability is given by

$$\begin{aligned}
&P\{\mathbf{P}(\bullet) \ is \ acceptable\} \\
&= P\{\|\mathbf{s}^*\|_\infty < \|\mathbf{s}\|_\infty\} \\
&= 1 - P\{\|\mathbf{s}^*\|_\infty \geq \|\mathbf{s}\|_\infty\} \\
&= 1 - \sum_{k=\|\mathbf{s}\|_\infty}^{M} P\{\|\mathbf{s}^*\|_\infty = k\}.
\end{aligned}$$

Let the incident Λ_1 be the occurrence of $\|\mathbf{s}^*\|_\infty = k$ and the incident Λ_2 the occurrence of $\exists j, \ s_j^* = k$. If Λ_1 occurs, then Λ_2 must occur; not vice-versa. It means that the cardinality of Λ_1 is not greater than that of Λ_2 and furthermore,

$$P\left\{\|\mathbf{s}^*\|_\infty = k\right\} \leq P\left\{\exists j, \ s_j^* = k\right\}.$$

On the other hand, apparently,

$$P\left\{\exists j, \ s_j^* = k + 1\right\} \leq P\left\{\exists j, \ s_j^* = k\right\}.$$

Thus,

$$
\begin{aligned}
&\mathrm{P}\left\{\mathbf{P}\left(\bullet\right) \ \ is\ \ acceptable\right\} \\
&\geq 1 - \textstyle\sum_{k=\|\mathbf{s}\|_\infty}^{M} \mathrm{P}\left\{\exists\, j,\ s_j^* = k\right\} \\
&\geq 1 - (M - \|\mathbf{s}\|_\infty + 1)\,\mathrm{P}\left\{\exists\, j,\ s_j^* = \|\mathbf{s}\|_\infty\right\}.
\end{aligned}
$$

Let $p = \frac{\|\mathbf{s}\|_1}{MN} = \frac{s}{MN}$.

$$
\mathrm{P}\left\{\exists\, j,\ s_j^* = \|\mathbf{s}\|_\infty\right\} = \binom{M}{\|\mathbf{s}\|_\infty} p^{\|\mathbf{s}\|_\infty}(1 - p)^{M - \|\mathbf{s}\|_\infty}.
$$

Due to the fact that $s \ll MN$ and then p is very small, we have

$$
\binom{M}{\|\mathbf{s}\|_\infty} p^{\|\mathbf{s}\|_\infty}(1 - p)^{M - \|\mathbf{s}\|_\infty} \approx \frac{\lambda^{\|\mathbf{s}\|_\infty}}{(\|\mathbf{s}\|_\infty)!} e^{-\lambda},
$$

where $\lambda = pM = \frac{s}{N}$. $(\|\mathbf{s}\|_\infty)!$ can be calculated according to Stirling's approximation as follows

$$
(\|\mathbf{s}\|_\infty)! \approx \sqrt{2\pi \|\mathbf{s}\|_\infty} \left(\frac{\|\mathbf{s}\|_\infty}{e}\right)^{\|\mathbf{s}\|_\infty},
$$

hence,

$$
(M - \|\mathbf{s}\|_\infty + 1)\,\mathrm{P}\left\{\exists\, j,\ s_j^* = \|\mathbf{s}\|_\infty\right\}
$$

$$
\begin{aligned}
&\approx \frac{(M - \|\mathbf{s}\|_\infty + 1)\lambda^{\|\mathbf{s}\|_\infty} e^{-\lambda}}{\sqrt{2\pi \|\mathbf{s}\|_\infty}(\|\mathbf{s}\|_\infty)^{\|\mathbf{s}\|_\infty} e^{-\|\mathbf{s}\|_\infty}} \\
&= \frac{(M - \|\mathbf{s}\|_\infty + 1)}{\sqrt{2\pi \|\mathbf{s}\|_\infty}} \left(\frac{\lambda}{\|\mathbf{s}\|_\infty}\right)^{\|\mathbf{s}\|_\infty} e^{\|\mathbf{s}\|_\infty - \lambda} \\
&= \frac{(M - \sigma\lceil \frac{s}{N}\rceil + 1)}{\sqrt{2\pi\sigma\cdot\lceil \frac{s}{N}\rceil}} \left(\frac{\lambda}{\sigma\cdot\lceil \frac{s}{N}\rceil}\right)^{\sigma\cdot\lceil \frac{s}{N}\rceil} e^{\sigma\cdot\lceil \frac{s}{N}\rceil - \lambda}, \\
&\leq \frac{(M - \sigma\lceil \frac{s}{N}\rceil + 1)}{\sqrt{2\pi\sigma\cdot\lceil \frac{s}{N}\rceil}} \left(\frac{\lceil \frac{s}{N}\rceil}{\sigma\cdot\lceil \frac{s}{N}\rceil}\right)^{\sigma\cdot\lceil \frac{s}{N}\rceil} e^{\sigma\cdot\lceil \frac{s}{N}\rceil - \lambda} \\
&= \frac{(M - \sigma\lceil \frac{s}{N}\rceil + 1)}{\sqrt{2\pi\sigma\cdot\lceil \frac{s}{N}\rceil}} \left(\frac{1}{\sigma}\right)^{\sigma\cdot\lceil \frac{s}{N}\rceil} e^{\sigma\cdot\lceil \frac{s}{N}\rceil - \lambda} \\
&= C \cdot \left(\frac{e}{\sigma}\right)^{\|\mathbf{s}\|_\infty}
\end{aligned}
$$

where the constant $C = \frac{(M - \sigma\lceil \frac{s}{N}\rceil + 1)}{\sqrt{2\pi\sigma\cdot\lceil \frac{s}{N}\rceil}e^\lambda} < M$. Generally, as long as $\|\mathbf{s}\|_\infty$ is large enough, it can guarantee $C \cdot \left(\frac{e}{\sigma}\right)^{\|\mathbf{s}\|_\infty} < 1$. With the increase of $\|\mathbf{s}\|_\infty$, the value of $C \cdot \left(\frac{e}{\sigma}\right)^{\|\mathbf{s}\|_\infty}$ decreases exponentially and converges to zero. Therefore, the random permutation is able to be an acceptable permutation with overwhelming probability.

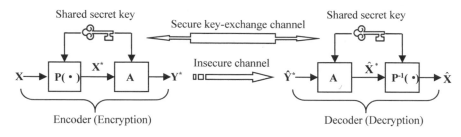

Fig. 3.7 A schematic diagram of the proposed approach

3.2.2 *Embedding Random Permutation in Parallel Compressed Sensing*

A block diagram of our approach is depicted in Fig. 3.7. The encoding process is mainly comprised of two steps, random permutation and Gaussian measurement. A 2D signal $\mathbf{X} \in \mathbb{R}^{M \times N}$ is firstly reshaped into a 1D signal $\{x(i)\}_{i=1}^{MN}$, which is then performed by random permutation. The permuted signal $\{x^*(i)\}_{i=1}^{MN}$ is converted back to the 2D format $\mathbf{X}^* \in \mathbb{R}^{M \times N}$. After the permutation, the signal $\mathbf{X}^*$ is sampled column by column using the same i.i.d. zero mean Gaussian measurement matrix $\boldsymbol{\Phi}$, i.e., $\mathbf{Y}^*[j] = \boldsymbol{\Phi}\mathbf{X}^*[j]$, where $\mathbf{Y}^* \in \mathbb{R}^{K \times N}$ and $\mathbf{X}^*[j]$ represents the jth column of $\mathbf{X}^*$. In the decoding phase, $\hat{\mathbf{X}}^*$ can be recovered from the received $\hat{\mathbf{Y}}^*$ and is then processed by the reverse permutation to derive the signal $\hat{\mathbf{X}}$ of interest, as shown in Fig. 3.7.

In what follows, we investigate the security of the proposed scheme embedding random permutation in PCS. Assume that Alice sends an encrypted message $\mathbf{Y}^* = \boldsymbol{\Phi}\mathbf{P}(\mathbf{X}) = \boldsymbol{\Phi}\mathbf{X}^*$ to Bob, who decrypts the message by solving the following convex optimization problem

$$\min \left\| \mathbf{X}^*[j] \right\|_1 \ s.t. \ \mathbf{Y}^*[j] = \boldsymbol{\Phi}\mathbf{X}^*[j], \ j \in [1, N] \tag{3.1}$$

and so $\mathbf{X} = \mathbf{P}^{-1}(\mathbf{X}^*)$. An eavesdropper, Eve, attempts to recover the plaintext $\mathbf{X}$ or the encryption keys $\boldsymbol{\Phi}$ and $\mathbf{P}(\bullet)$ after intercepting the ciphertext $\mathbf{Y}^*$.

3.2.2.1 Asymptotic Spherical Secrecy

Considering Shannon's definition of perfect secrecy that the probability of a message conditioned on the cryptogram is equal to the a priori probability of the message, the proposed scheme does not achieve perfect secrecy, as stated in Lemma 2.

Lemma 2 *Let X be a random variable, whose probability is $P_X(\mathbf{X}) > 0$, $\forall \mathbf{X} \in \mathbb{R}^{M \times N}$, and $\boldsymbol{\Phi}$ be a $K \times M$ measurement matrix. With respect to the encryption model $Y = \boldsymbol{\Phi}\mathbf{P}(X)$, we have $I(X; Y) > 0$, and so perfect secrecy is not achieved.*

Proof We prove this lemma by contradiction. Apparently, $I(X; Y) > 0$ if and only if X and Y are statistically independent. In the context of $X = \mathbf{0}$, $Y = \mathbf{\Phi}P(X) = \mathbf{\Phi}P(\mathbf{0}) = \mathbf{\Phi} \cdot \mathbf{0} = \mathbf{0}$ and so $P_{Y|X}(Y = \mathbf{0}|X = \mathbf{0}) = 1$. On the other hand, only $\mathbf{X}$ in the null space of $\bar{\mathbf{\Phi}}$ which is a new transform $\bar{\mathbf{\Phi}} = \mathbf{\Phi}P(\bullet)$ are mapped to $Y = \mathbf{0}$; whereas, we have $P_Y(Y = \mathbf{0}) < 1$ due to the assumption that $P_X(\mathbf{X}) > 0$, $\forall \mathbf{X} \in \mathbb{R}^{M \times N}$. As a result, we conclude that $P_{Y|X}(Y = \mathbf{0}|X = \mathbf{0}) \neq P_Y(Y = \mathbf{0})$, meaning that X and Y are statistically dependent.

By the RIP, $\mathbf{Y}$ provides information about the norm of $\mathbf{X}$. The fact that the l_2-norm of a vector can be considered as its energy has been utilized by Cambareri et al. [74] in introducing the notion of asymptotic spherical secrecy for CS encoding in which the measurement matrix serves as a key.

Definition 2 ([74]) (*Asymptotic spherical secrecy*).
Let $\mathbf{x}^{(n)} = (x_0, x_1, \ldots, x_n) \in \mathbb{R}^n$ be a plaintext sequence of increasing length n and $\mathbf{y}^{(n)}$ be the corresponding ciphertext sequence. Assume that the power of the plaintext is positive and finite, i.e.,

$$W_{\mathbf{x}} = \lim_{n \to \infty} \frac{1}{n} \sum_{k=1}^{n} x_k^2, \quad 0 < W_{\mathbf{x}} < +\infty. \tag{3.2}$$

A cryptosystem is said to have asymptotic spherical secrecy if

$$f_{Y^{(n)}|X^{(n)}}(\mathbf{y}, \mathbf{x}) \xrightarrow{D} f_{Y^{(n)}|W_{\mathbf{x}}}(\mathbf{y}), \tag{3.3}$$

where $\xrightarrow{D}$ denotes the convergence in distribution as $n \to \infty$.

This definition implies that it is impossible for Eve to infer the plaintext $\mathbf{x}$ but its power from the statistical properties of the random measurements $\mathbf{y}$. Although not achieving perfect secrecy, the proposed scheme satisfies asymptotic spherical secrecy.

Theorem 2 (Asymptotic spherical secrecy of the proposed scheme)
Let
(1) $\mathbf{X}^{(n)} = (X_{ij}) \in \mathbb{R}^{M \times N}$ *be a bounded-value plaintext with power* $0 < W_{\mathbf{X}} < +\infty$, *defined as* $W_{\mathbf{X}} = \lim_{n \to \infty} \frac{1}{n} \sum_{i=1}^{M} \sum_{j=1}^{N} X_{ij}^2$, *where* $n = MN$;
(2) $\mathbf{X}^{*(n)} = \mathbf{P}(\mathbf{X}^{(n)}) = (X_{ij}^*) \in \mathbb{R}^{M \times N}$
with power $W_{\mathbf{X}^*} = \lim_{n \to \infty} \frac{1}{n} \sum_{i=1}^{M} \sum_{j=1}^{N} (X_{ij}^*)^2$;
(3) $\mathbf{Y}^{(n)} = (Y_{ij}) \in \mathbb{R}^{K \times M}$ *be the corresponding ciphertext, where* $Y_{ij} = \sum_{k=1}^{M} \Phi_{ik} X_{kj}^*$. *As* $n \to \infty$, *we have*

$$Y_{ij} \xrightarrow{D} N(0, MW_{\mathbf{X}}/K). \tag{3.4}$$

Proof Permutation does not affect the power and thus $W_{\mathbf{X}^*} = W_{\mathbf{X}}$. After the random permutation, the energy is approximately uniformly distributed to each column of

$\mathbf{X}^{*(n)}$. In other words, the power of each column converges to that of the whole plaintext in distribution, i.e.,

$$
\begin{aligned}
\frac{1}{M} \sum_{k=1}^{M} \left(X_{ij}^{*} \right)^{2} &\xrightarrow{D} \lim_{n \to \infty} \frac{1}{n} \sum_{i=1}^{M} \sum_{j=1}^{N} \left(X_{ij}^{*} \right)^{2} \\
&= \lim_{n \to \infty} \frac{1}{n} \sum_{i=1}^{M} \sum_{j=1}^{N} X_{ij}^{2} = W_{\mathbf{X}}.
\end{aligned}
\tag{3.5}
$$

We calculate

$$
\begin{aligned}
\mathbf{E}\left[Y_{ij}^{2} \right] &= \mathbf{E}\left[\left(\sum_{k=1}^{M} \Phi_{ik} X_{kj}^{*} \right)^{2} \right] \\
&= \frac{1}{K} \sum_{k=1}^{M} \left(X_{kj}^{*} \right)^{2} \xrightarrow{D} \frac{M}{K} W_{\mathbf{X}},
\end{aligned}
\tag{3.6}
$$

thereby yielding the result stated in Theorem 2.

3.2.3 The Realization of Chaos

In order to validate the feasibility of the proposed encryption models at first, we utilize chaos to implement the proposed scheme embedding random permutation in PCS. The whole process is under the control of the skew tent chaos map with four keys, μ, $z(0)$, μ' and $z'(0)$.

3.2.3.1 Generate the Permutation Order

There are a number of classic methods in realizing the permutation operation $\mathbf{P}(\bullet)$ from one or more keys using chaos, some of which are stated as follows:

Straightforward transform. Use 2D chaotic transforms such as Arnold map to directly project the indices of the 2D signal, e.g., [75].

Matrix rotation. Employ chaotic sequence to construct the rotation matrix acting on 1D signal, e.g., [2].

Index sorting. Sort the chaotic sequence to generate the index matrix, applying to the indices of the 1D signal, e.g., [76].

Here, we apply the method "index sorting" to perform the permutation. According to [76], a permutation sequence is produced using the skew tent map by the following steps:

a. Set the keys μ and $z(0)$ to iterate the skew tent map $MN + m$ times, then discard the first m values to get rid of the transient effect.

b. Sort the remaining MN values $\{z(i)\}_{i=m+1}^{m+MN}$ to obtain $\{\bar{z}(i)\}_{i=m+1}^{m+MN}$.

c. Search the values of $\{z(i)\}_{i=m+1}^{m+MN}$ in $\{\bar{z}(i)\}_{i=m+1}^{m+MN}$, and store the corresponding indices $\{Index(i)\}_{i=1}^{MN}$.

Apparently, $\{Index\,(i)\}_{i=1}^{MN}$ indicates an order of the integers from 1 to MN. The above steps have been widely used to generate the permutation order in image encryption schemes. However, the complexity $\mathcal{O}\left(n^2 \log n\right)$ is high. A novel algorithm, whose complexity is only $\mathcal{O}\left(n \log n\right)$, was designed in [11]. The procedures are:

    a. Initialize a flag sequence $\{flag\,(k)\}_{k=1}^{MN}$ and a permutation sequence $\{Index\,(k)\}_{k=1}^{MN}$ to 0 and set $i = 1$.

    b. Calculate $z\,(k+1) = T\,[z\,(k)\,;\mu]$ and $\chi = \lceil MN \times z\,(k+1)\rceil$.

    c. If $flag\,(\chi) = 0$, then set $Index\,(i) = \chi$, $flag\,(\chi) = 1$ and $i = i + 1$; otherwise, go to Step b.

    d. If $i < MN$, go to Step b.

3.2.3.2 Construct the Measurement Matrix

Following the idea of [77], the chaotic measurement matrix is constructed by the following steps:

    a. Define the chaotic sequence

$$\Delta\left(d, k, \mu', z'\,(0)\right) := \left\{z'\,(n + i \times d)\right\}_{i=0}^{k}, \tag{3.7}$$

which is extracted from the chaotic sequence generated by the skew tent map with sampling distance d and keys μ' and $z'\,(0)$.

    b. Introduce a new transform

$$\{\vartheta\,(k)\}_{k=0}^{KM-1} = \left\{-2 \times \Delta\left(d, k, \mu', z'\,(0)\right) + 1\right\}\big|_{k=KM-1}. \tag{3.8}$$

    c. Create the measurement matrix column by column using the sequence $\{\vartheta\,(k)\}_{k=0}^{KM-1}$, as given by

$$\Phi = \sqrt{\frac{2}{K}}\begin{pmatrix} \vartheta\,(0) & \cdots & \vartheta\,(KM - K) \\ \vdots & \ddots & \vdots \\ \vartheta\,(K - 1) & \cdots & \vartheta\,(KM - 1) \end{pmatrix} \tag{3.9}$$

where the scalar $\sqrt{2/K}$ is used for normalization.

3.2.3.3 Computational Secrecy

Cryptosystems relying on computation-secrecy such as RSA are practical and widely used. In contrast to information theoretic secrecy which is an ideal encryption requirement, computational secrecy allows the ciphertext possessing complete or partial plaintext information, which is common. This ensures that for Eve to recover the plaintext from the ciphertext without the correct key is equivalent to solving a computational problem that is assumed to be extremely difficult (e.g., NP-hard). In the proposed scheme, $\mathbf{X}$ is a 2D sparse signal with sparsity $\mathbf{s}$. If a wrong key μ, z (0), μ' or z' (0), which is almost identical to the correct key, is used by Eve in attempting to recover $\mathbf{X}$, the result is unsuccessful due to the high key sensitivity. Moreover, the unsuccessful recovery of the signal using a wrong key can also be justified by the following theorem.

Theorem 3   ([44]) *Let $\mathbf{\Phi}$ and $\tilde{\mathbf{\Phi}}$ be $K \times M$ Gaussian matrices with entries generated by different keys. Let $\mathbf{x}$ be s-sparse and $\mathbf{y} = \mathbf{\Phi}x$. When $\tilde{s} \geq s + 1$, the l_0 or l_1 optimization used will yield an $\tilde{s}$-sparse solution with probability one.*

On the contrary, once an s-sparse solution is obtained using some keys, Eve easily realizes that it must be the correct key. Computational secrecy can be achieved if Eve is computationally bounded; otherwise, the cryptanalysis will succeed. However, in practical applications, the keys should be at least 2^{64} bits and are updated periodically to resist brute-force attack.

Every communication requires altering the session keys, which can be securely transmitted. For instance, they are encrypted by public-key encryption algorithms such as RSA. Apparently, it can resist the potential attacks including known-plaintext attack, chosen-plaintext attack and chosen-ciphertext attack. It is also impossible for the attacker to cryptanalyze the proposed approach using cipher-only attack, since analyzing the encoded data, having a smaller amount than the original data, to retrieve the original data without knowing the secret measurement matrix is an NP-hard question.

3.2.4   Simulation

For simulation purpose, an image can be considered as a 2D signal, which is sparsified by 2D discrete cosine transform (DCT2) to obtain a 2D sparse signal $\mathbf{X}$. The best s-term approximation is acquired by keeping the s largest DCT2 coefficients and setting the remaining to zeros. Random permutation and Gaussian matrix are generated by using MATLAB code. Four images of size 512×512, Peppers, Lena, Boat and Baboon, are used in the simulations. The basis pursuit algorithm in the CVX optimization toolbox [78] is employed to realize the PCS reconstruction. Apart from the basis pursuit, other reconstruction algorithms can also be used. The reconstruction performance is evaluated by peak signal-to-noise ratio (PSNR).

3.2.4.1   Compressibility

The encoded (or encrypted) image can have various sizes depending on the compression ratio (CR), i.e., the ratio of the number of measurements to the total number of entries in the DCT2 coefficient matrix. Figure 3.8 shows four encoded images of the original Peppers image corresponding to $CR = 0.8, 0.6, 0.4, 0.2$. In order to inves-

(a)   **(b)**   **(c)**   **(d)**

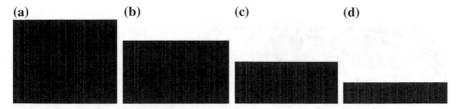

Fig. 3.8 Four encoded images at different CRs. **a** $CR = 0.8$; **b** $CR = 0.6$; **c** $CR = 0.4$; **d** $CR = 0.2$

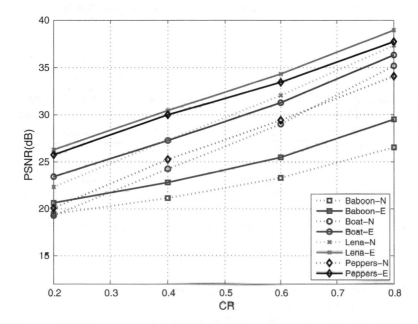

Fig. 3.9 PSNR versus CR for different images with/without random permutation encryption

tigate the effect of encryption on CR, we plot PSNR versus CR for different images with/without encryption in Fig. 3.9, where "E" represents introducing random permutation encryption while "N" means not introducing, which refers to the case that a 2D sparse signal is sampled column by column using the same measurement matrix drawn from Gaussian ensembles. As observed from Fig. 3.9, random permutation helps to improve the PSNR of all images by around 2–6 dB at the same CR. In other words, at the same PSNR, random permutation encryption makes CR smaller. This is due to the fact that random permutation can relax the RIP for 2D sparse signals with high probability in PCS, as justified by Theorem 1.

(a)   (b)   (c)   (d)

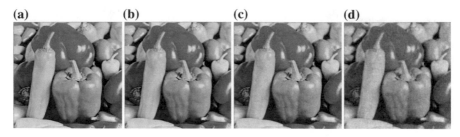

Fig. 3.10 Four decoded images corresponding to the four encoded images in Fig. 3.8 at various CRs. **a** $CR = 0.8$; **b** $CR = 0.6$; **c** $CR = 0.4$; **d** $CR = 0.2$

3.2.4.2 Robustness

Introducing encryption into PCS makes it still possess high reconstruction robustness, even for a small amount of encoded data. This can be visually verified by the four decoded images shown in Fig. 3.10. The decoded (or decrypted) images contain most of the visual information of the original images, even at $CR = 0.2$. A significant requirement in the transmission process is the robustness of a coding system (or cryptosystem) against imperfection such as additive white Gaussian noise (AWGN) and cropping attack (CA). These two capabilities are quantified in Table 3.1 for the Peppers image. In particular, the encoded images at different CRs are affected by these imperfections and the PSNRs of the corresponding decoded images are calculated. The AWGN yields zero-mean normal distribution with variance 1 while the cropping attack cuts one-eighth of the encoded image at the upper left corner. Observing Γ_3 and Γ_4, or Γ_5 and Γ_6 from Table 3.1, we can see that at a channel with both AWGN and CA, random permutation encryption improves the PSNR at the same CR. By individually comparing the variation trends of $\Gamma_2 - \Gamma_1$, $\Gamma_4 - \Gamma_3$ and $\Gamma_6 - \Gamma_5$, the tendency is that the smaller the CR, the greater the improvement. In addition, vertically contrasting these three rows of data reveals that PSNR improvements are similar with and without AWGN (or CA). Thus, we come to the conclusion that the proposed approach possesses a strong robustness against AWGN and CA. It is worth mentioning that similar results are obtained using other images. In addition, to test the sensitivity of the four keys, a tiny perturbation of 10^{-16} is added, respectively, and the decoded images are depicted in Fig. 3.11. Their indistinguishability justifies the high key sensitivity of the proposed approach. In fact, this is guaranteed by the inherent property of chaos, i.e., high sensitivity to initial conditions. The key space is at least 2^{64}.

Table 3.1 PSNR of different settings at various CRs

CR	0.2	0.4	0.6	0.8
PCS-N(=Γ_1)	20.0800	25.2575	29.4243	34.0949
PCS-E(=Γ_2)	25.7507	30.0071	33.4539	37.7431
$\Gamma_2 - \Gamma_1$	5.6707	4.7496	4.0296	3.6482
PCS-AWGN-N(=Γ_3)	20.0899	25.2382	29.3072	33.0931
PCS-AWGN-E(=Γ_4)	25.7405	29.9143	33.0547	35.8942
$\Gamma_4 - \Gamma_3$	5.6506	4.6761	3.7475	2.8011
PCS-CA-N(=Γ_5)	19.3763	24.6003	28.8997	33.5010
PCS-CA-E(=Γ_6)	25.1834	29.4669	33.0240	37.3011
$\Gamma_6 - \Gamma_5$	5.8071	4.8666	4.1243	3.8001

(a)   (b)   (c)   (d)

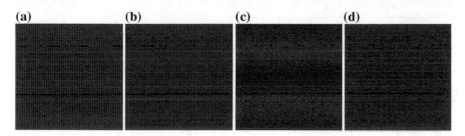

Fig. 3.11 The test of key sensitivity ($CR = 0.2$). **a** $z(0) = 0.33 + 10^{-16}$; **b** $\mu = 0.63 + 10^{-16}$; **c** $z'(0) = 0.73 + 10^{-16}$; **d** $\mu' = 0.28 + 10^{-16}$

3.3 The Involvement of Image Processing Techniques

The existing Block Compressive Sensing (BCS) based image ciphers adopted the same sampling rate for all the blocks, which may lead to the desirable result that after subsampling, significant blocks lose some more-useful information while insignificant blocks still retain some less-useful information. Motivated by this observation, we propose a scalable encryption framework (SEF) based on BCS together with a Sobel Edge Detector (SED) and Cascade Chaotic Maps (CCM). Our work is firstly dedicated to the design of two new fusion techniques, chaos-based structurally random matrices and chaos-based random convolution and subsampling. The basic idea is to divide an image into some blocks with an equal size and then diagnose their respective significance with the help of SED. For significant block encryption, chaos-based structurally random matrix is applied to significant blocks whereas chaos-based random convolution and subsampling are responsible for the remaining insignificant ones. In comparison with the BCS based image ciphers, the SEF takes lightweight subsampling and severe sensitivity encryption for the significant blocks and severe subsampling and lightweight robustness encryption for the insignificant ones in parallel, thus better protecting significant image regions.

3.3.1 *Preliminaries*

3.3.1.1 SED

Edge detection is a fundamental tool in image processing and computer vision, particularly in the areas of feature detection and feature extraction. It aims at identifying the location of sharp intensity transitions in an image. The SED implements a 2D spatial gradient measurement on an input gray-scale image and is suitable for the approximate absolute gradient magnitude at each point by detecting regions of high spatial gradient [79]. In SED, the image is convolved with a pair of 3×3 convolution masks, which respond maximally to edges running vertically and horizontally relative to the pixel grid one mask for each of the two perpendicular orientations. One mask is simply rotated by $90°$ to acquire the other. These masks can produce separate measurements of the gradient component in each orientation when applied separately to the input image. By combining the measurements, the absolute magnitude of the gradient at each point and the orientation of that gradient can be found.

3.3.1.2 CCM

A general chaotic system called CCM was introduced in [80] by leveraging two 1D maps as seed maps to generate numerous new chaotic maps, which match the cascade structure in electronic circuits. For example, consider two common 1D maps, e.g., the Logistic map and the Tent map respectively, mathematically written as

$$z(i+1) = \alpha z(i)[1 - z(i)], \tag{3.10}$$

and

$$z(i+1) = \begin{cases} \beta z(i), & z(i) < 0.5 \\ \beta[1 - z(i)], & z(i) \geq 0.5, \end{cases} \tag{3.11}$$

where the parameters $\alpha \in [3.57, 4]$ and $\beta \in (1, 2]$. They can be cascaded into a Logistic-Tent map or a Tent-Logistic map. The Tent-Logistic map is defined as

$$z(i+1) = \begin{cases} \alpha\beta z(i)[1 - \beta z(i)], & z(i) < 0.5 \\ \alpha\beta[1 - z(i)]\{1 - \beta[1 - z(i)]\}, & z(i) \geq 0.5. \end{cases} \tag{3.12}$$

It has been demonstrated that this type of newly created CCM is more unpredictable and has better chaotic performance in comparison with individual seed maps [80].

3.3.2 Two Fusion Technologies

This section introduces two new fusion technologies, CSRM and CRCS, which are the basis for the encryption design of the SEF.

3.3.2.1 CSRM

Structurally random matrix (SRM) for CS was proposed by Do et al. [81], which simultaneously has the following features: optimal or near optimal sensing performance, universality, low complexity and hard/optics implementation friendliness. The SRM is comprised of three multiplication matrices:

$$\Phi = \sqrt{m/n}\mathbf{DMR}, \tag{3.13}$$

where $\mathbf{R} \in \mathbb{R}^{n \times n}$ is either a uniform random permutation matrix or a diagonal random matrix whose diagonal entries are Bernoulli random variables; $\mathbf{M} \in \mathbb{R}^{n \times n}$ represents an orthonormal matrix selected among popular fast computable transforms such as Fast Fourier Transform (FFT), DCT, and Walsh-Hadamard Transform (WHT); $\mathbf{D} \in \mathbb{R}^{m \times n}$ is a subsampling operator that selects a random subset of rows frow the matrix $\mathbf{MR}$. The most outstanding characteristic of SRM is the pre-randomization step for the purpose of scrambling the structure of the signal, which converts the sensing signal into a white noise-like one to achieve universally incoherent sensing.

The permutation and diffusion are two basic properties that an ideal cryptosystem should have [82]. The permutation-diffusion has been widely applied in chaos-based image cryptography. The permutation process aims at confusing the pixel positions while the diffusion operation is for the purpose of altering pixel values. In the following, we propose a new permutation-diffusion-permutation framework using SRM. For an input 1D vector, $\mathbf{R}$, which is used as a uniform random permutation matrix, rearranges the positions of this vector and $\mathbf{M}$ performs the diffusion to change the element values for the permuted vector. While, $\mathbf{D}$, as a subsampling operator, can be implemented such that another uniform random permutation matrix $\mathbf{R}' \in \mathbb{R}^{n \times n}$ confuses the diffused vector followed by a truncated manipulation of the front elements. Correspondingly, the new SRM is described as:

$$\left[1 : m^{(1)}\right] \odot \Phi^{(1)} = \left[1 : m^{(1)}\right] \odot \sqrt{m^{(1)}/n}\mathbf{R}'\mathbf{MR} \tag{3.14}$$

where $\left[1 : m^{(1)}\right] \odot$ means a truncation of the first $m^{(1)}$ terms. In the proposed permutation-diffusion-permutation framework, two permutations can be generated by chaotic maps. In other words, the SRM shown in Eq. 3.14 is under the control of a chaotic map, which forms a new fusion technique called CSRM.

3.3.2.2 CRCS

Random convolution and subsampling (RCS) for CS was presented by Romberg [83], in which RCS as a universally efficient CS paradigm was demonstrated. A signal $\mathbf{x} \in \mathbb{R}^n$ is circularly convolved with a pulse $\mathbf{h} \in \mathbb{R}^n$ and then subsampled. The convolution of $\mathbf{x}$ and $\mathbf{h}$ can be algebraically written as $\mathbf{Hx}$, where

$$\mathbf{H} = n^{-1/2}\mathbf{F}^*\mathbf{\Lambda F}, \tag{3.15}$$

in which $\mathbf{F}$ is the discrete Fourier matrix and $\mathbf{\Lambda} = \mathrm{diag}\{\sigma_1, \sigma_2, \ldots, \sigma_n\}$ with σ_l being unit magnitude complex numbers with random phases as follows: $\sigma_1 = \pm 1$ with equal probability; $\sigma_{n/2+1} = \pm 1$ with equal probability; $\sigma_l = \exp(j\theta_l)$, where $2 \le l \le n/2$ and $\theta_l \in [0, 2\pi]$ yields uniform distribution; $\sigma_l = \sigma_{n-l+2}^*$, i.e., the conjugate of σ_{n-l+2}, where $n/2 + 2 \le l < n$. After random convolution, the subsampling is performed by observing entries of $\mathbf{Hx}$ at a small number of randomly chosen locations.

The DRPE proposed by Refregier and Javidi [51] is the most basic structure in optical image ciphers. In analogy to the DRPE, the RCS can also be implemented by a double random structure, including random phase and random permutation, followed by a truncation. The unified measurement matrix is formulated as

$$\left[1 : m^{(2)}\right] \odot \mathbf{\Phi}^{(2)} = \left[1 : m^{(2)}\right] \odot \mathbf{R}''n^{-1/2}\mathbf{F}^*\mathbf{\Lambda F}, \tag{3.16}$$

where $\mathbf{\Lambda}$ and $\mathbf{R}''$ represent random phase and random permutation, respectively. Likewise, they can be offered by chaotic maps, which brings a benefit of not destroying the robustness of the double random structure but keeping the initial value sensitivity of chaos. This fusion technique is a shorthand for CRCS.

Note that, the above two fusion techniques, CSRM and CRCS, are evolved from two classic encryption techniques of permutation-diffusion and DRPE, respectively, as the fundamentals of two mainstreams, chaos-based image cipher and optical image cipher, in the image security field.

3.3.3 Scalable Encryption Framework

The basic idea of the proposed SEF is that an image is firstly divided into some blocks with an equal size and then, with the help of SED, the significance of each block is diagnosed. For significant block encryption, CSRM is applied whereas CRCS is responsible for the remaining insignificant ones. The whole workflow is illustrated in Fig. 3.12.

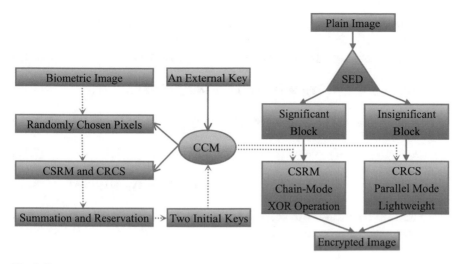

Fig. 3.12 Block diagram of the proposed SEF

3.3.3.1 Biometrics and Subsampling Motivated Key Generation

In recent years, biometrics, such as fingerprints, retinas, palm prints, facial structures, is used to generate keys in the encryption process. It brings above an advantage that key management becomes unique, untraceable and secure. For example, Bhatnagar and Wu presented such a biometrics inspired multimedia encryption scheme based on dual parameter FrFT [84], which captures a biometric image of owner and generates biometrically encoded bitstream. Our study makes use of the CSRM and CRCS theories to generate two keys, respectively, with a biometric image and an external initial key shared by the decryption side.

CSRM and CRCS Measurement Matrices Construction.
The CSRM and CRCS measurement matrices are generated by invoking Algorithms 2 and 3, respectively, as follows.

a. Let ek be an external key, assign $IK \leftarrow ek$ and invoke Algorithm 2 to generate $\mathbf{R} \leftarrow \mathbf{R}_p$ and an new value IK.
b. Invoke Algorithm 2 with IK to generate $\mathbf{R}' \leftarrow \mathbf{R}_p$ and an new value IK.
c. Invoke Algorithm 2 with IK to generate $\mathbf{R}'' \leftarrow \mathbf{R}_p$ and an new value IK.
d. Invoke Algorithm 3 with IK to generate $\mathbf{\Lambda}$ and an new value IK.

Key Generation.
Key generation is performed by using a biometric image and subsampling.

a. Let $\mathbf{B}$ be a biometric image. Assume that it contains ζ ($\zeta > n$) pixels in all. With IK, invoke Algorithm 4 to generate an randomly chosen vector $\mathbf{v}_1 \leftarrow \mathbf{v}$ of length n and an new value IK. Moreover, invoke Algorithm 4 with IK to generate another randomly chosen vector $\mathbf{v}_2 \leftarrow \mathbf{v}$ of length n.

b. Exploit 1/3 subsampling for $\mathbf{v}_1$ and $\mathbf{v}_2$ by using CSRM and CRCS as follows:

$$\mathbf{y}_{\mathbf{v}_1} = [1:n/3] \odot \sqrt{m^{(1)}/n} \mathbf{R}'\mathbf{MR}\mathbf{v}_1, \tag{3.17}$$

$$\mathbf{y}_{\mathbf{v}_2} = [1:n/3] \odot \mathbf{R}'' n^{-1/2} \mathbf{F}^* \mathbf{\Lambda} \mathbf{F} \mathbf{v}_2. \tag{3.18}$$

c. Summarize $\mathbf{y}_{\mathbf{v}_1}$ and $\mathbf{y}_{\mathbf{v}_2}$, respectively, and only reserve the fractional parts given as

$$K_1 = real\left(sum\left(\mathbf{y}_{\mathbf{v}_1}\right) - floor\left(sum\left(\mathbf{y}_{\mathbf{v}_1}\right)\right)\right), \tag{3.19}$$

$$K_2 = real\left(sum\left(\mathbf{y}_{\mathbf{v}_2}\right) - floor\left(sum\left(\mathbf{y}_{\mathbf{v}_2}\right)\right)\right), \tag{3.20}$$

where $real\,(\cdot)$ means to truncate the real part. These two numbers are used as two keys for the encryption framework.

Algorithm 1 CCM Iteration.

Input: An initial value z, α, β.
Output: A new value z.
1: **If** $z < 0.5$ then
2:     $z \leftarrow \alpha\beta z\,[1 - \beta z]$;
3: **Else**
4:     $z \leftarrow \alpha\beta\,(1 - z)\,[1 - \beta\,(1 - z)]$;
5: **End**

Algorithm 2 Random Permutation Matrix Generation.

Input: An initial value IK, α, β.
Output: An random permutation matrix $\mathbf{R}_p$ of size $n \times n$ and IK.
1: With an input value $z \leftarrow IK$, invoke Algorithm 1 τ times to obtain an new initial iteration value z, where $\tau \geq 1000$ to improve the initial value sensitivity.
2: Set an identical permutation matrix $\mathbf{R}_p \leftarrow \mathbf{I}_{n \times n}$.
3: **For** $i \leftarrow n$ down to 2
4:     Invoke Algorithm 1 once;
5:     Set $k \leftarrow floor\left(z \times 2^{14} \bmod i\right) + 1$;
6:     Swap the ith and kth rows in $\mathbf{R}_p$;
7: **End**
8: $IK \leftarrow z$.

3.3.3.2 Encryption Framework

The encryption framework is comprised of three modules, including significance recognition, significant block encryption and insignificant block encryption.

Algorithm 3 Random Phase Matrix Generation.

Input: An initial value IK, α, β.
Output: An random phase matrix $\mathbf{\Lambda} = \mathrm{diag}\{\sigma_1, \sigma_2, \ldots, \sigma_n\}$ and IK.
1: With an input value $z \leftarrow IK$, invoke Algorithm 1 τ ($\tau \geq 1000$) times to obtain an new initial iteration value z.
2: Set a zero matrix $\mathbf{\Lambda} \leftarrow \mathbf{0}_{n \times n}$.
3: **For** $l \leftarrow 2$ to $n/2$
4:     Invoke Algorithm 1 once;
5:     $\sigma_l \leftarrow \exp(2\pi j \times z)$;
6:     $\sigma_{n-l+2} \leftarrow \exp(-2\pi j \times z)$;
7: **End**
8: Invoke Algorithm 1 once;
9: If $z < 0.5$, then $\sigma_1 \leftarrow +1$; else $\sigma_1 \leftarrow -1$.
10: Invoke Algorithm 1 once;
11: If $z < 0.5$, then $\sigma_{n/2+1} \leftarrow +1$; else $\sigma_{n/2+1} \leftarrow -1$.
12: $IK \leftarrow z$.

Algorithm 4 Randomly Chosen Vector Generation.

Input: An initial value IK, α, β, $\mathbf{B}$.
Output: An randomly chosen vector $\mathbf{v}$ and IK.
1: With an input value $z \leftarrow IK$, invoke Algorithm 1 τ ($\tau \geq 1000$) times to obtain an new initial iteration value z.
2: Set a zero vector $\mathbf{v} = 0_{1 \times n}$.
3: Set $cnt \leftarrow 0$, $i \leftarrow numel(\mathbf{B})$.
4: **While** $cnt < n$
5:     Invoke Algorithm 1 once;
6:     Set $k \leftarrow floor\left(z \times 2^{14} \bmod i\right) + 1$;
7:     $v(cnt) \leftarrow B(k)$ and then delete the element $B(k)$ in $\mathbf{B}$;
8:     $i \leftarrow i - 1$, $cnt \leftarrow cnt + 1$;
9: **End**
10: $IK \leftarrow z$.

Significance Recognition.

Different regions in an image often provide different levels of significance. In general, the significance of edge regions which contain contour features is higher than that of smooth ones. For the purpose of recognizing significance, edge detection technique, which is a fundamental and essential pre-processing step in image segmentation and computer vision applications, can be introduced to identify the significant contour features. In our study, the edge detector called SED is employed.

    a. Input a gray-scale plain image $\mathbf{P}$ of size $\sqrt{N} \times \sqrt{N}$.

    b. Exploit SED for $\mathbf{P}$ to output the corresponding detection image $\mathbf{P}'$, which is referred to as the *association* of $\mathbf{P}$. The $\mathbf{P}'$ is a binary image consisting of "1" and "0", which reflect the detected and undetected pixels, respectively.

    c. Partition $\mathbf{P}$ and $\mathbf{P}'$ into non-overlapping blocks in the same way, each of which contains $\sqrt{n} \times \sqrt{n}$ pixels. Every original block p_i, $i = 1, 2, \ldots, N/n$, from $\mathbf{P}$ has a corresponding associated block p_i' from $\mathbf{P}'$. There are N/n pairs of blocks in total.

d. Count the number μ_i of the detected pixels for each associated block and then obtain the significant degree $d_i = \mu_i/n$.

e. Set the threshold level $T \in [0, 1]$ and generate a binary significant vector (BSV) from all the d_i, in which $BSV_i = 1$, i.e., $d_i \geq T$ means that the p_i is significant whereas $BSV_i = 0$ is opposite. (The above method is derived from the previous work [15]). Let the numbers of significant and insignificant blocks be N_{sig} and N_{ins}, respectively, and let the significant and insignificant block sets be $S^{sig} = \left\{ p_1^{sig}, p_2^{sig}, \ldots, p_{N_{sig}}^{sig} \right\}$ and $S^{ins} = \left\{ p_1^{ins}, p_2^{ins}, \ldots, p_{N_{ins}}^{ins} \right\}$, respectively. Note that, $N_{sig} + N_{ins} = N/n$.

Significant Block Encryption.

For security enhancement purpose, two features are introduced, including the chain mode that the present block encryption result will affect the next block encryption and a new permutation-diffusion-permutation-diffusion architecture that CSRM with permutation-diffusion-permutation architecture is followed by XOR diffusion operation.

a. Rearrange each significant block p_1^{sig} into a vector $\mathbf{p}_1^{sig}$ according to the raster mode. Invoke Algorithm 5 with K_1 to generate $\mathbf{R}$ and $\mathbf{R}'$. Then $\mathbf{p}_1^{sig}$ is subsampled by using CSRM as

$$\mathbf{p}_{s,1}^{sig} = \left[1 : m^{(1)} \right] \odot \sqrt{m^{(1)}/n} \mathbf{R}' \mathbf{M} \mathbf{R} \mathbf{p}_1^{sig}. \tag{3.21}$$

b. Invoke Algorithm 6 with K_1 to generate a keystream $\mathbf{k}_1$ of length $m^{(1)}$. XORing $\mathbf{p}_{s,1}^{sig}$ and $\mathbf{k}_1$ element by element with regard to integer part as

$$\mathbf{c}_1^{sig} = \left(integer \left(\mathbf{p}_{s,1}^{sig} \right) \oplus integer \left(abs \left(\mathbf{k}_1 \right) \right) \right) + fraction \left(\mathbf{p}_{s,1}^{sig} \right). \tag{3.22}$$

c. Prior to the next significant block encryption, K_1 is updated by using $\mathbf{c}_1^{sig}$ as

$$K_1 = \left(sum \left(\mathbf{c}_1^{sig} \right) - floor \left(sum \left(\mathbf{c}_1^{sig} \right) \right) + K_1 \right) \bmod 1. \tag{3.23}$$

d. Manipulate the next significant block p_2^{sig} in the similar steps (a–b) and then update K_1 in the similar step (c), until the last significant block is encrypted.

Algorithm 5 R and R′ Generation.

Input: K_1, α, β.
Output: R and R′.
1: $IK \leftarrow K_1$, invoke Algorithm 2 to acquire: $\mathbf{R} \leftarrow \mathbf{R}_p$ and IK.
2: Invoke Algorithm 2 to acquire $\mathbf{R}' \leftarrow \mathbf{R}_p$.

Algorithm 6 $\mathbf{k}_1$ Generation.

Input: K_1, α, β.
Output: $\mathbf{k}_1$.
1: $IK \leftarrow K_1$, invoke Algorithm 1 $\tau + n$ times and combine the last n outputs to be a vector $\mathbf{v}_{\mathbf{k}_1}$.
2: $\mathbf{v}_{\mathbf{k}_1} \leftarrow floor\left(\mathbf{v}_{\mathbf{k}_1} \times 2^{14}\right) \bmod L$, where L represents the image gray level.
3: $\mathbf{k}_1 \leftarrow \left[1 : m^{(1)}\right] \odot \sqrt{m^{(1)}/n}\mathbf{R}'\mathbf{MR}\mathbf{v}_{\mathbf{k}_1}$.

Insignificant Block Encryption.
Unlike significant block encryption, insignificant block encryption is performed by two simple steps:

    a. Invoke Algorithm 7 with K_2 to generate Λ and $\mathbf{R}''$.

    b. Subsample each block separately by CRCS as

$$\mathbf{c}_i^{ins} = \left[1 : m^{(2)}\right] \odot \mathbf{R}''n^{-1/2}\mathbf{F}^*\Lambda\mathbf{F}\mathbf{p}_i^{ins}. \tag{3.24}$$

Algorithm 7 Λ and $\mathbf{R}''$ Generation.

Input: K_2, α, β.
Output: Λ and $\mathbf{R}''$.
1: $IK \leftarrow K_2$, invoke Algorithm 3 to generate Λ and IK.
2: Invoke Algorithm 2 to generate $\mathbf{R}''$.

3.3.3.3 Decryption Framework

The decryption framework proceeds as follows.

K_1 and K_2 Generation.
This step is the same as the phase of biometrics and subsampling motivated key generation in the encryption side.

Significant Block Decryption.
It has the following steps:

    a. Invoke Algorithm 6 with K_1 to generate a keystream $\mathbf{k}_1$ of length $m^{(1)}$. XORing $\mathbf{c}_1^{sig}$ and $\mathbf{k}_1$ element by element with regard to integer part as

$$\mathbf{p}_{s,1}^{sig} = \left(integer\left(\mathbf{c}_1^{sig}\right) \oplus integer\left(abs\left(\mathbf{k}_1\right)\right)\right) + fraction\left(\mathbf{c}_1^{sig}\right). \tag{3.25}$$

    b. Invoke Algorithm 5 with K_1 to generate $\mathbf{R}$ and $\mathbf{R}'$. Apply a reconstruction algorithm to obtain $\mathbf{p}_1^{sig,dec}$ and then rearrange it into a block $p_1^{sig,dec}$.

    c. Update K_1 as Eq. 3.23.

    d. Manipulate the next significant ciphertext $\mathbf{c}_2^{sig}$ in the similar steps (a–b) and then update K_1 in the similar step (c), until the last significant ciphertext is decrypted.

Insignificant Block Decryption.
It has only two steps:
    a. Invoke Algorithm 7 with K_2 to generate Λ and $\mathbf{R}''$.
    b. Apply a reconstruction algorithm to obtain $\mathbf{p}_i^{ins,dec}$ and then rearrange them into blocks $p_i^{ins,dec}$.

Assembling Blocks.
Package the significant block ciphertext $p_i^{sig,dec}$ and the insignificant ciphertext $p_i^{ins,dec}$ into a complete image with the help of BSV.

3.3.4   Further Discussions

3.3.4.1   Sensitive and Robust Encryption

The proposed SEF is comprised of two different types of encryption models including sensitive encryption and robust encryption, which correspond to the significant block encryption and the insignificant block encryption, respectively. The former adopts the CSRM technique and chain-mode with XOR operation. The CSRM with the architecture of permutation-diffusion-permutation is suitable for protecting important information. The chain-mode with XOR operation means that the present significant block encryption will affect the next one. The loss or a small modification of the present significant block ciphertext will result in that the next significant block ciphertext cannot be correctly decrypted, since the present ciphertext is responsible for updating the initial key of the next significant block decryption and the key sensitivity of CCM causes a great change. As a result, it is a sensitive encryption model. The latter only exploits a subsampling operation by the CRCS technique for each insignificant block. The CRCS as a variant of DRPE has the robustness of CS itself and optical encryption. It is robust against noise attack and cropping attack. Therefore, the latter is a robust encryption model.

3.3.4.2   Lightweight and Severe Subsampling

The important data often take up a much smaller proportion than the unimportant data in an image, since the size of contour regions is often much less than that of smooth regions. That is, the number of significant blocks is far smaller than that of insignificant ones. In fact, this can come true easily through adjusting the threshold level. A natural idea is to sample more data for significant blocks but less data for insignificant ones. Indeed, significant blocks are fully maintained and insignificant ones are severely sub-sampled. On the whole, one should exert a lightweight subsampling and sensitive encryption on significant blocks and a severe subsampling and robust encryption on insignificant ones.

3.3.4.3 Parallelization and Progressive Transmission

Despite the larger number of insignificant blocks, it can work efficiently in a parallel computing environment to carry out subsampling and reconstruction, due to the fact that each block is individually processed. Furthermore, the proposed SEF can bring about progressive transmission because of the global block encryption mode. In the decryption side, once a significant or insignificant block is recovered, it can be previewed. The decryption of all the significant blocks helps preview a rough sketch of an image and the decryption of each insignificant block supplements some specific texture details. The ciphertext contains both the significant block ciphertext and the insignificant block ciphertext. The significant ciphertext needs to be transmitted over a noiseless channel and the insignificant ciphertext is also transmitted over a noisy channel.

3.3.4.4 Double Protection

The security of the proposed SEF depends on a biometric image and three parameters (ek, α, β). None of them will fail in decryption. Biometrics has the key management advantages in acting as the keys, like the biometrics-inspired encryption techniques [84–86], but they may be duplicated or embezzled with the progress of biometric techniques. Fortunately, the parameters (ek, α, β) provide another layer of protection and can be updated once. On the other hand, only the parameter keys, like the traditional chaotic image ciphers [1, 3, 5, 7, 9, 11, 13, 87], have no superiorities of key management. As a consequence, the proposed SEF incorporates the strengths of both the biometrics-inspired encryption techniques and the traditional chaotic image ciphers, delivering the effect of double protection.

3.3.5 Experimental Analysis

A fingerprint image of size 220×244 and four gray-scale images of size 512×512, as shown in Fig. 3.13, are chosen to test the performance of the proposed SEF in MATLAB platform. There are seven parameters $\left(ek, \alpha, \beta, \text{T}, n, m^{(1)}, m^{(2)}\right)$, where the first five ones without specification are set as $ek = 0.7589, \alpha = 4, \beta = 2, \text{T} = 0.2$ and $n = 64$, respectively. The $\mathbf{M}$ is selected as DCT. The GPSR [88] algorithm is used for reconstruction and the DCT is employed as the sparsifying basis. It is very crucial for an ideal image cryptosystem to be immune to some common attacks such as vision security analysis, sensitivity analysis or robustness analysis, and differential attack analysis. The followings give a detailed investigation of these attacks to the proposed SEF.

(a)   (b)   (c)   (d)   (e)

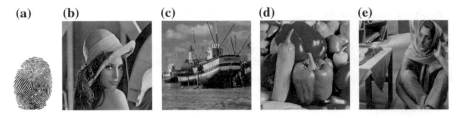

Fig. 3.13 A fingerprint image and four gray-scale images used for test

3.3.5.1   Vision Security Analysis

The $m^{(1)}$ and $m^{(2)}$ are set as $m^{(1)} = n$ and $m^{(2)} = 0.2n$, respectively. The SED results are depicted in Fig. 3.14a–d. After performing SEF, the ciphertexts are unintelligible from human visual system, as verified in Fig. 3.14e–l. In spite of the same $m^{(1)}$ and $m^{(2)}$, different images suffer from inconsonant compression ratio (CR), as the same threshold level T distinguishes the unfixed numbers of significant and insignificant blocks. The CRs of four images are 0.2244, 0.2405, 0.1940 and 0.2807, respectively. It can be shown from the decryption results in Fig. 3.14m–p that with such low CRs, the decrypted images are still visible to some extent. Meanwhile, contrasting textures of four images, as shown in Fig. 3.14a–d, with corresponding CRs deduces that for an image, one can see that the more complicated the texture, the higher the CR. The CR is adaptive about texture when keeping the same experimental setup. As a matter of fact, this deduction can be explained by the fact that the more complicated texture implies the more number of significant blocks and the smaller number of insignificant blocks. The weaker sampling degree of the former versus the latter (e.g., 1 vs 0.2) will enlarge the CR.

3.3.5.2   Sensitivity and Robustness Analysis

In general, an image cipher is either sensitive or robust. However, the proposed SEF is a combination of sensitive encryption and robust encryption, corresponding to the significant block encryption and the insignificant block encryption, respectively. The whole process, whether it is the biometrics and subsampling motivated key generation or the encryption framework, is under the control of CCM. This CCM-driven mechanism reserves the sensitivity of key generation and encryption due to its intrinsic initial value sensitivity. For key sensitivity test purpose, a slightly altered key $K_1' = K_1 + 10^{-10}$ is used to decrypt the significant block encryption results generated by K_1, and then four decrypted images are given in Fig. 3.15. The value of T is set as 0.05 for visual effect enhancement. In addition, the robustness is tested by adding AWGN into the insignificant block ciphertext and Fig. 3.16 lists the decrypted images. From Figs. 3.15 and 3.16, the significant blocks (contour regions) are blurred

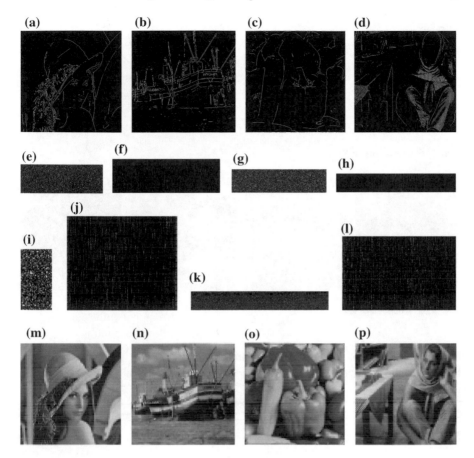

Fig. 3.14 The SED, encryption and decryption results: **a–d**. The SED results; **e, g, i, k**. The significant block encrypted images for Lena, Boat, Peppers and Barbara, respectively; **f, h, j, l**. The insignificant block encrypted images for Lena, Boat, Peppers and Barbara, respectively; **m–p**. The decrypted images

(comparing to Fig. 3.14m–p) and the insignificant blocks (smooth regions) possess a certain degree of visibility, which well verify the sensitivity and robustness.

3.3.5.3   Histogram Analysis

The main purpose of histogram analysis is to distinguish whether the ciphertexts of different images yield a similar distribution. Observing Fig. 3.17, we have the knowledge that the histograms of the significant block and insignificant block ciphertexts are completely different from that of the original images. Comparing the ciphertext histograms finds that all the four significant block ciphertexts present an approximate Gaussian distribution and four insignificant block ciphertexts follow an approximate

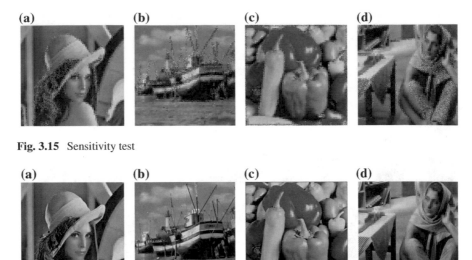

Fig. 3.15 Sensitivity test

Fig. 3.16 Robustness test

Rayleigh distribution. This excellent statistical property can prevent information leakage, since no attackers can infer the original image distribution from the cipher-text.

3.3.5.4 Comparative Analysis

In comparison with the BCS based image ciphers, which adopted the same subsampling way and encryption mode for each block, the SEF takes lightweight subsampling and severe encryption for the significant blocks and severe subsampling and lightweight encryption for the insignificant blocks in parallel. For comparative analysis, the benchmark cipher is to perform the same subsampling and encryption for all the blocks using the insignificant block encryption way while other conditions keep unchanged, as the proposed SEF. Moreover, in the benchmark cipher, the CRs are set as 0.2244, 0.2405, 0.1940 and 0.2807 for four images, respectively. Then, the reconstructed images are shown in Fig. 3.18. By contrasting Figs. 3.14m–p and 3.18, it is concluded that the SEF has better visual effect in significant blocks than the benchmark cipher does. Thus, the SEF can well protect significant image regions. If the SEF and the benchmark cipher have consistent visual effects, the SEF will have a stronger compression result, which reduces power consumption during transmission and storage space at the receiving end.

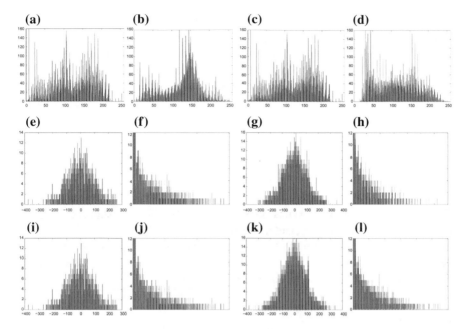

Fig. 3.17 Histogram test: **a–d**. Histgrams of four original images; **e, g, i, k**. Histograms of the significant block encrypted images for Lena, Boat, Peppers and Barbara, respectively; **f, h, j, l**. Histograms of the insignificant block encrypted images for Lena, Boat, Peppers and Barbara, respectively

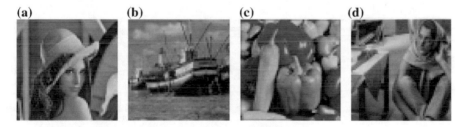

Fig. 3.18 Four decrypted images for the benchmark cipher

3.4 Double Protection Mechanism for Compressive Sensing

Some pioneering works have investigated embedding cryptographic properties in CS in a way similar to one-time pad symmetric cipher. We tackles the problem of constructing a CS-based symmetric cipher under the key reuse circumstance, i.e., the cipher is resistant to common attacks even a fixed measurement matrix is used multiple times. To this end, we suggest a bi-level protected CS (BLP-CS) model which makes use of the advantage of measurement matrix construction without RIP. Specifically, two kinds of artificial basis mismatch techniques are investigated to construct key-related sparsifying bases. The encoding process of BLP-CS is simply a random linear projection, which is the same as the basic CS model. However, decoding the linear measurements requires knowledge of both the key-dependent sensing matrix and its sparsifying basis. The proposed model is exemplified by sampling images as a joint data acquisition and protection layer for resource-limited wireless sensors. Simulation results and numerical analyses have justified that the new model can be applied in circumstances where the measurement matrix can be re-used.

3.4.1 Bi-Level Protection Model

The block diagram of this model is shown in Fig. 3.19, where we suggest using key-dependent sensing matrix, $\mathbf{A}_K$, and secret-related sparsifying basis, $\mathbf{\Psi}_K$, to determine the measurement matrix $\mathbf{\Phi} = \mathbf{A}_K \mathbf{\Psi}_K^{-1}$. The measurement matrix $\mathbf{\Phi}$ does not satisfy the RIP requirement, while the key-dependent sensing matrix $\mathbf{A}_K$ itself is a RIP matrix. The sampling procedure can be expressed as

$$\mathbf{y} = \mathbf{\Phi}\mathbf{x} = \mathbf{A}_K \mathbf{\Psi}_K^{-1}(\mathbf{\Psi}_K \mathbf{s}) = \mathbf{A}_K \mathbf{s}. \tag{3.26}$$

To correctly decode (decrypt) $\mathbf{y}$, a legitimate user should first derive $\mathbf{A}_K$ and $\mathbf{\Psi}_K$ from the key scheduling process and then refer to the following two-step reconstruction

$$\min \|\mathbf{s}\|_1 \text{ subject to } \mathbf{y} = \mathbf{\Phi}\mathbf{x} = \mathbf{A}_K \mathbf{s}, \mathbf{x} = \mathbf{\Psi}_K \mathbf{s}. \tag{3.27}$$

or equivalently

$$\min \|\mathbf{\Psi}_K^{-1}\mathbf{x}\|_1 \text{ subject to } \mathbf{y} = \mathbf{\Phi}\mathbf{x}. \tag{3.28}$$

To fulfill the security requirement, the remaining task is to design two matrices $\mathbf{A}_K$ and $\mathbf{\Psi}_K$ satisfying:

Fig. 3.19 Block diagram of BLP-CS

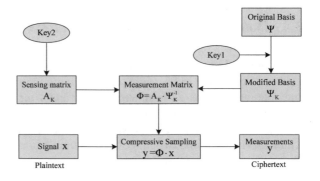

- RULE a: $\mathbf{A}_K$ is a key-related matrix satisfy RIP
- RULE b: $\mathbf{\Psi}_K$ is a key-related sparsifying basis
- RULE c: $\mathbf{A}_K \mathbf{\Psi}_K^{-1}$ is a structural non-RIP matrix

The work of designing a RIP matrix is trivial since it is already clear that Gaussian/Bernoulli [42] and structurally random matrices [81] are competent for this task with overwhelming probability. Therefore, we focus our attention on the designing of $\mathbf{\Psi}_K$ in the following discussions. It is worth mentioning that the work of designing $\mathbf{\Psi}_K$ satisfying RULE b (also known as transform encryption) is very popular in the filed of multimedia encryption, examples can be found in [89–91]. However, the work of designing $\mathbf{A}_K$ and $\mathbf{\Psi}_K$ satisfying RULE c is totally new.

3.4.2 Type I Secret Basis

The first type of secret basis that drawn our attention is the parameterized construction of some familiar transform, such as parameterized DWT [90, 92] and directional DCT [89, 93]. Here, we present a parameterized transform based on FrFT as an example.

The use of FrFT for security purpose can be dated back to year 2000, when Unnikrishnan et al. [94] suggested to use FrFT for DRPE instead of the ordinary Fourier transform [95], in order to benefit from its extra degrees of freedom provided by the fractional orders. Generally speaking, performing an order α FrFT on a signal can be viewed as a rotation operation on the time-frequency or space-frequency distribution at an angle α. Though FrFT is very popular in optics for its easy implementation, it is not preferred in digital world since complex numbers always cause extra computational load.

To this end, Venturini et al. proposed a method to construct Reality-Preserving FrFT of arbitrary order [96]. Here, we deduce the Reality-Preserving Fractional Cosine Transform (RPFrCT) by the virtue of their method. Denote the DCT [97] of size $n \times n$ by

$$\mathbf{C} = \frac{1}{\sqrt{n}} \epsilon_l \cos(2\pi \frac{(2i+1)l}{4n}) \tag{3.29}$$

where $i = 0 \sim n - 1$, $l = 0 \sim n - 1$, $\epsilon_0 = 1$ and $\epsilon_l = \sqrt{2}$ for $l > 0$. The unitary property of $\mathbf{C}$ assures that it can be diagonalized as

$$\mathbf{C} = \mathbf{U\Lambda U}^*, \tag{3.30}$$

where $\mathbf{U} = \{\mathbf{u}_i\}_{i=1}^n$ is composed of n orthonormal eigenvectors, i.e., $\mathbf{u}_m^* \mathbf{u}_i = \delta_{mi}$ and $\mathbf{\Lambda} = \mathrm{diag}(\lambda_1, \ldots, \lambda_i, \ldots, \lambda_n)$ with $\lambda_i = \exp(j\varphi_i)$. Replace λ_i with its α-th power λ_i^α in Eq. (3.30), we can express the Discrete Fractional Cosine Transform (DFrCT) matrix $\mathbf{C}_\alpha$ of order α in the compact form

$$\mathbf{C}_\alpha = \mathbf{U\Lambda}^\alpha \mathbf{U}^*. \tag{3.31}$$

Having defined $\mathbf{C}_\alpha$, we can derive the RPFrCT matrix $\mathbf{R}_\alpha$ as follows:

- For any real signal $\mathbf{x} = \{x_l\}_{l=1}^M$ of length M (M is even), construct a complex signal of length $M/2$ by

$$\widetilde{\mathbf{x}} = \{x_1 + jx_{M/2+1}, x_2 + jx_{M/2+2}, \ldots, x_{M/2} + jx_M\}. \qquad (3.32)$$

- Compute $\widetilde{\mathbf{y}} = \mathbf{B}_\alpha \widetilde{\mathbf{x}}$, where $\mathbf{B}_\alpha$ is a DFrCT matrix of size $(M/2 \times M/2)$, namely, $\mathbf{B}_\alpha = \mathbf{C}_{\alpha, M/2}$.
- Determine the RPFrCT matrix $\mathbf{R}_\alpha$ by

$$\begin{aligned}
\mathbf{y} &= (\Re(\widetilde{\mathbf{y}}), \Im(\widetilde{\mathbf{y}}))^T \\
&= \begin{pmatrix} \Re(\mathbf{B}_\alpha)\Re(\widetilde{\mathbf{x}}) - \Im(\mathbf{B}_\alpha)\Im(\widetilde{\mathbf{x}}) \\ \Im(\mathbf{B}_\alpha)\Re(\widetilde{\mathbf{x}}) + \Re(\mathbf{B}_\alpha)\Im(\widetilde{\mathbf{x}}) \end{pmatrix} \\
&= \begin{pmatrix} \Re(\mathbf{B}_\alpha) & -\Im(\mathbf{B}_\alpha) \\ \Im(\mathbf{B}_\alpha) & \Re(\mathbf{B}_\alpha) \end{pmatrix} \cdot \begin{pmatrix} \Re(\widetilde{\mathbf{x}}) \\ \Im(\widetilde{\mathbf{x}}) \end{pmatrix} \\
&= \mathbf{R}_\alpha \mathbf{x}.
\end{aligned}$$

From the construction process listed above, we can conclude that $\mathbf{R}_\alpha$ is orthogonal, reality preserving and periodic. Then, the Reality-Preserving Fractional Cosine Transform of a digital image $\mathbf{X}$ is given by

$$\mathbf{S} = \mathbf{R}_\alpha \mathbf{X} \mathbf{R}_\beta^T, \qquad (3.33)$$

where $(\cdot)^T$ represents the transpose operator, α and β are the orders of the Fractional Cosine Transform along x and y directions, respectively. Equivalently, we can express this formula as

$$(\mathbf{S}) = \mathbf{\Psi}^{-1}(\mathbf{X}), \qquad (3.34)$$

where $\mathbf{\Psi}^{-1} = \mathbf{\Psi}^T = (\mathbf{R}_\beta \otimes \mathbf{R}_\alpha)$. To study the sparsifying capability of the proposed parameterized basis, we carried out experiments on digital images at different fractional orders α and β by using the best s-term approximation, i.e., keep the s largest coefficients and set the remaining ones to zero. The recovered result of RPFrCT is compared with that of DCT2 using the ratio between their PSNRs. As expected, the sparsifying capability of RPFrCT raises when α or β increases, as shown in Fig. 3.20. When $\alpha, \beta \in (0.9, 1]$, the sparsifying capability of RPFrCT is comparable to that of DCT2. It is worth mentioning that a similar sparsifying capability was also observed when this transform is applied to 1D signals [96].

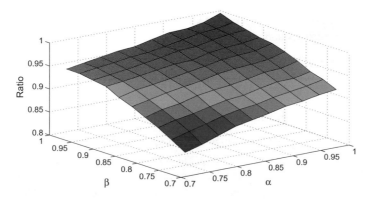

Fig. 3.20 Comparison between the recovery result of RPFrCT and DCT2 using the best s-term approximation at different fractional orders

3.4.3   Type II Secret Basis

We have demonstrated a technique for parameterized sparsifying basis construction, where the free parameter can be used as the secret key in the BLP-CS model. In this way, the resultant basis satisfies RULE b. However, it still suffers from the same CPA since it fails to meet RULE c. In the subsequent discussions, we propose three kind of operations on an existing basis to make it fulfill RULE c. We start the deviation by defining equivalent sparsifying bases.

Definition 3   Two basis matrices, $\boldsymbol{\Psi}$ and $\boldsymbol{\Psi}'$ are equivalent sparsifying bases if $\mathbf{x} = \boldsymbol{\Psi}\mathbf{s} = \boldsymbol{\Psi}'\mathbf{s}'$, $\|\mathbf{s}\|_0 = \|\mathbf{s}'\|_0 = k$ holds for any signal $\mathbf{x}$.

Property 1 $\boldsymbol{\Psi}'$ and $\boldsymbol{\Psi}$ are equivalent sparsifying bases if

$$\boldsymbol{\Psi}' = \mathbb{F}_1(\boldsymbol{\Psi}) = (d_1\boldsymbol{\Psi}_1, d_2\boldsymbol{\Psi}_2, \ldots, d_j\boldsymbol{\Psi}_j, \ldots, d_M\boldsymbol{\Psi}_M),$$

where $\{d_j\}_{j=1}^M$ are non-zero constants and ψ_j is the j-th column of $\boldsymbol{\Psi}$.

Proof   Set $s'_j = \frac{1}{d_j}s_j$ and we have $\|\mathbf{s}\|_0 = \|\mathbf{s}'\|_0$.

We demonstrate that we are able to construct a non-RIP measurement matrix satisfying RULE c. Assume $\boldsymbol{\Psi}$ is an orthonormal basis and set

$$\boldsymbol{\Psi}' = \boldsymbol{\Psi}\mathbf{D},$$

where $\mathbf{D} = \mathrm{diag}(1/d_1, 1/d_2, \ldots, 1/d_M)$ and $\{d_j\}_{j=1}^M$ are positive integers drawn from certain distribution independently. Let $\mathbf{A}$ denote a Gaussian matrix with i.i.d. entries and calculate $\boldsymbol{\Phi}$ as

$$\boldsymbol{\Phi} = \mathbf{A}(\boldsymbol{\Psi}\mathbf{D})^{-1}, = \mathbf{A}\mathbf{D}^{-1}\boldsymbol{\Psi}^T.$$

Once again, the effect of $\mathbf{\Psi}^T$ can be viewed as a rotation of $\mathbf{AD}^{-1}$ in a M dimensional space, which is energy preserving. By construction, $\mathbf{\Phi}$ is a non-RIP matrix.

Property 2 $\mathbf{\Psi}'$ and $\mathbf{\Psi}$ are equivalent sparsifying bases if

$$\mathbf{\Psi}' = \mathbb{F}_2(\mathbf{\Psi}) = \mathbf{\Psi}\mathbf{P},$$

where $\mathbf{P}$ is a random permutation matrix.

Proof Since $\mathbf{\Psi s} = \mathbf{\Psi}(\mathbf{PP}^T)\mathbf{s} = \mathbf{\Psi}'(\mathbf{P}^T\mathbf{s}) = \mathbf{\Psi}'\mathbf{s}'$, $\|\mathbf{s}'\|_0 = \|\mathbf{P}^T\mathbf{s}\|_0 = \|\mathbf{s}\|_0$.

In the 1D case, this property implies that random scrambling does not cause any loss of the sparsity level of any given signal. In the 2D case, it helps to uniform the column (or row) sparsity level and thus flavors a PCS reconstruction technique, which will be exemplified in Sect. 3.4.5.

In addition, if we know or partially know that supp($\mathbf{s}$) is localized in a certain k-dimensional subspace rather than uniformly distributed in $\mathbb{R}^M$, we can embed more secrets into the sparsifying basis, as stated in Property 3. Here we assume that $\mathbf{\Psi}$ is an orthonormal sparsifying basis for simplicity.

Property 3 $\mathbf{\Psi}'$ and $\mathbf{\Psi}$ are equivalent sparsifying bases if

$$\mathbf{\Psi}' = \mathbb{F}_3(\mathbf{\Psi}) = (\psi_1, \ldots, \psi_{j-1}, a\psi_j + b\psi_k, \psi_{j+1}, \ldots, \psi_M), \tag{3.35}$$

where a, b are non-zero constants and $j, k \in$ supp($\mathbf{s}$) or $j, k \notin$ supp($\mathbf{s}$).

Proof Since $\mathbf{\Psi}$ is orthonormal, $s_j = (\mathbf{\Psi}_j, \mathbf{x}) = \mathbf{\Psi}_j^T \mathbf{x}$ and we know $s_j = 0$ when $j \notin$ supp($\mathbf{s}$). Then the proof for $j, k \notin$ supp($\mathbf{s}$) is trivial. For $j, k \in$ supp($\mathbf{s}$), set $\mathbf{s}' = (s_1', s_2', \ldots, s_j', \ldots, s_k', \ldots, s_M')^T$ with

$$s_i' = \begin{cases} s_i/a & \text{if } i = j, \\ s_i - s_j b/a & \text{if } i = k, \\ s_i & \text{otherwise.} \end{cases} \tag{3.36}$$

Then we have

$$\begin{aligned} \mathbf{x} &= \mathbf{\Psi s} \\ &= \sum_{\substack{i=1 \\ i \neq j,k}}^M s_i\psi_i + s_j\psi_j + s_k\psi_k \\ &= \sum_{\substack{i=1 \\ i \neq j,k}}^M s_i\psi_i + \frac{s_j}{a}(a\psi_j + b\psi_k) + (s_k - \frac{bs_j}{a})\psi_k \\ &= \mathbf{\Psi}'\mathbf{s}' \end{aligned} \tag{3.37}$$

By Eq. (3.36), we conclude that $\|\mathbf{s}'\|_0 = \|\mathbf{s}\|_0$, hence completes the proof.

Obviously, the operator $\mathbb{F}_3(\cdot)$ can be applied to three or more columns as long as all of the chosen columns are either in supp($\mathbf{s}$) or not. Finally, we provide an example to further illustrate Property 3. The grayscale image "Lena" with size 512×512, as shown in Fig. 3.21a, is transformed using RPFrCT with orders $\alpha = 0.99$ and $\beta = 0.95$. Figure 3.21b shows the absolute value of the RPFrCT coefficients under

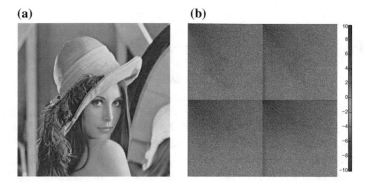

Fig. 3.21 **a** Original image "Lena"; **b** Energy distribution of RPFrCT coefficients of "Lena" using logarithm base

the logarithm base. It is clear that the energy of the RPFrCT coefficients matrix is localized, specifically, they are concentrated at the upper-left corner of the four sub-blocks. Thus, we can apply Property 3 to the RPFrCT basis $\boldsymbol{\Psi} = (\mathbf{R}_\beta \otimes \mathbf{R}_\alpha)^T$ accordingly. A similar effect can be observed in the parameterized DWT and DCT settings.

3.4.4 Discussions and Security Analysis

We have demonstrated the possibility of using BLP-CS as a joint data acquisition and protection model for MTS purpose. This section aims to compare the basic OTS CS cipher and BLP-CS cipher from the viewpoints of complexity and security.

3.4.4.1 Complexity

Suppose we have constructed a RPFrCT matrix $\mathbf{R}_\alpha$ with appropriate fractional order α, a $M \times 1$ signal $\mathbf{x}$ can be sparsified by $\mathbf{R}_\alpha \mathbf{x} = \mathbf{s}$. All the techniques on manipulating the sparsifying basis $\mathbf{R}_\alpha^T$ introduced in Sect. 3.4.3 can be unified to the following matrix notation,[1] i.e.,

$$\boldsymbol{\Psi}_K = \mathbf{R}_\alpha^T \mathbf{PDQ}, \tag{3.38}$$

where $\mathbf{D}$, $\mathbf{P}$ and $\mathbf{Q}$ are matrices determined by operators $\mathbb{F}_1$, $\mathbb{F}_2$ and $\mathbb{F}_3$, respectively. It worth mentioning that $\mathbf{x} = \boldsymbol{\Psi}_K \mathbf{s}' = \mathbf{R}_\alpha^T \mathbf{s}$ with $\|\mathbf{s}'\|_0 = \|\mathbf{s}\|_0$. Recall from Sect. 3.4.1, the encoding of BLP-CS is governed by

---

[1]We are aware of the fact that any parameterized orthonormal transform with good sparsifying capability can play the role of $\mathbf{R}_\alpha^T$.

$$\mathbf{y} = \mathbf{\Phi}\mathbf{x} = \mathbf{A}_K \mathbf{\Psi}_K^{-1}\mathbf{x}, \tag{3.39}$$

and the decoding should follow a two-step reconstruction, i.e.,

$$\min \|\mathbf{s}'\|_1 \text{ subject to } \mathbf{y} = \mathbf{\Phi}\mathbf{x} = \mathbf{A}_K \mathbf{s}', \mathbf{x} = \mathbf{\Psi}_K \mathbf{s}'. \tag{3.40}$$

Once a well-designed key schedule is given,[2] a trusted third party can produce $\mathbf{\Phi}$, $\mathbf{A}_K$ and $\mathbf{\Psi}_K$ faithfully and transmit them to the encoder and decoder. An alternative option is that the encoder and decoder produce their own matrix key on the air using the agreed key schedule from the same root key. We assume the OTS CS model also adopts the same matrix key generation process for a fair comparison.

We first take a look at the encoder side. For the former situation, where the matrix key is produced by the trusted party and then delivered to both the CS encoder and decoder, the encoding complexity of the BLP-CS model outperforms that of the OTS CS model since it does not bring extra communication cost once the key is set. For the later situation, the encoding complexity of the OTS CS model is lower than that of the BLP-CS model at the first glimpse due to the reason that the encoding process of the second model involves a matrix multiplication, i.e., $\mathbf{A}_K \mathbf{\Psi}_K^{-1}$, in the key generation process. Nevertheless, since the OTS CS system requires updating the measurement matrix for every sampling, the BLP-CS model outperforms OTS CS after sampling $(2f' + f)/f'$ times. Here, f and f' refer to the complexity of the matrix multiplication and the matrix key generation, respectively.

At the decoder side, the Moore-Penrose pseudoinverse of the sensing matrix $\mathbf{A}_K$ need to be calculated in every iteration of some l_1 optimization algorithms [78], for example, orthogonal matching pursuit [98]. The complexity of this operation dominates the overall complexity in CS reconstruction. As such, if some off-line techniques can be employed to calculate the pseudoinverse of $\mathbf{A}_K$, the complexity of the reconstruction can be largely reduced. For the OTS CS system, this is impossible since the measurement matrix is never re-used.

3.4.4.2   Security

1. *Brute-force and Ciphertext-only Attacks*. We employ the existing results presented in [44, 74] to show that the BLP-CS preserves most secrecy features of the OTS CS-based cipher under these two attacks. We first examine the case of brute-force attack, i.e., the attacker try to guess possible measurement matrices and use them for decoding. The l_0 or l_1 recovery governed by a wrong sensing matrix $\mathbf{A}_K$ will lead to an incorrect reconstruction with probability one. Thus the OTS CS-based cipher can guarantee computational secrecy if the key space is large enough to make systematic search of all the keys (sensing matrices) impossible. This result can be directly

---

[2]The design of an effective key scheduling process is not considered in this book since our concern is only the secrecy of CS paradigm. We also note that this is a common treatment for all the state-of-the-art works on this topic.

applied to our BLP-CS model. According Eqs. (3.39) and (3.40), we can conclude that BLP-CS is computationally strong even if the attacker can successfully retrieved the secret sparsifying basis $\mathbf{\Psi}_K$. In this concern, the transform encryption approach enhances the security level of the basic CS paradigm.

An interesting security feature of the OTS CS cryptosystem under ciphertext-only attack is the asymptotic spherical secrecy [74]. This type of secrecy states that any two different plaintexts (sparse signals to be sampled in this context) with equal power remain approximately indistinguishable from their measurement vectors when CS operates under the RIP framework. Alternatively, we can intercept this property as only the energy of the measurements carries information about the signal. A bird's-eye view of why this asymptotic spherical secrecy holds for the OTS CS cipher may refer to the definition of RIP, which states that the CS encoding should obey an energy-preserving guarantee. A theoretical proof about this property can be found in [74].

As we demonstrated in Eqs. (3.39) and (3.40), the proposed BLP-CS model works under the seemingly RIPless theory if one cannot determine $\mathbf{A}_K$ and $\mathbf{\Psi}_K$. Therefore, the energy-preserving constraint introduced by RIP is unapplicable to this setting. As such, we can conclude that the measurements (ciphertext) carries no information about the signal (plaintext) when a single ciphertext is observed. The BLP-CS and the OTS CS ciphers have the following major difference: when multiple ciphertexts are observed by the attacker, he is aware of the fact that two plaintexts must be similar if their corresponding ciphertexts are close to each other in the Euclidean space. This is caused by the multi-time usage of the same measurement matrix and the linear encoder. Surely the OTS CS cipher is more secure then the BLP-CS cipher from this point of view. Nevertheless, this is a favorable property that promotes the source coding gain from a system point-of-view [99]. This property also finds its way in privacy-preserving video surveillance systems [100]: assume the attacker happens to know some pairs of plaintext and ciphertext, such as static video scenes and their corresponding measurement vectors, and he want to retrieve privacy-sensitive data from a new intercepted ciphertext. After studying the Euclidean distance of the new ciphertext, he comes to realize that plaintext corresponding to the new ciphertext contains privacy-sensitive data. However, the decryption of this ciphertext requires full knowledge of the matrix key $\mathbf{A}_K$ and $\mathbf{\Phi}_K$. This leads to our discussion of resistance of the BLP-CS cipher with respect to plaintext attacks.

2. *Plaintext Attacks.* The data complexity of retrieving a general measurement matrix (the secret key) is M independent plaintexts and their corresponding ciphertexts in any basic CS-based cipher. If the used measurement matrix is Bernoulli, a single plaintext in the form $\mathbf{x} = (2^0, 2^1, \ldots, 2^M)^T$ and the corresponding ciphertext can be utilized to recover the Bernoulli measurement matrix completely.[3] Based on these knowledge, investigating the resistance of the OTS CS cryptosystem is a trivial work. We hereby focus on the BLP-CS cipher. Referring to Eq. (3.39), the attacker can retrieve $\mathbf{\Phi}$ from M independent plaintext-ciphertext pairs. By construction, $\mathbf{\Phi}$

---

[3]One can imagine the role of a $\{+1, -1\}$ matrix as that of a $\{0, 1\}$ matrix, the proof can be found in [101]. A vector composed by $\{0, 1\}$ can be recovered from the inner product of this vector and $\mathbf{x}$.

is a non-RIP matrix. Thus a straightforward use Φ in the l_1 optimization problem is not applicable. Considering that the l_0 optimization problem is NP-hard [102], the attacker tries to decompose Φ with the form $\Phi = \mathbf{EF}$, with the constraint that entries of $\mathbf{E}$ should observe certain kind of distribution (Gaussian or Bernoulli). In particular, $\mathbf{F}$ is the product of an elementary matrix and an orthonormal matrix.

If the decomposition is unique or the possible number of decompositions is very limited, i.e., polynomial function of M, the attacker can determine the matrix key $\mathbf{A}_K$ and $\mathbf{\Psi}_K^{-1}$ and the BLP-CS cryptosystem is regarded as fail to resist plaintext attacks. To summarize, we conclude that the number of decompositions should be at least $\mathcal{O}(M!)$, thus making the search for the true one inconclusive. The conclusion is based on the simple fact $\mathbf{EF} = (\mathbf{EP})(\mathbf{P}^T \mathbf{F})$, where $\mathbf{P}$ is a $M \times M$ random permutation matrix. As we can see, distribution of all the entries of $(\mathbf{EP})$ is exactly the same as that of $\mathbf{E}$ and $\mathbf{P}^T$ represents elementary row operation on $\mathbf{F}$. As such, the attacker cannot distinguish the decomposition result $\mathbf{E}$ and $\mathbf{F}$ from $(\mathbf{EP})$ and $(\mathbf{P}^T \mathbf{F})$.

3.4.5 BLP-CS for Digital Images

In this section, the proposed BLP-CS model is applied as a joint data acquisition and protection layer for digital images. The aim is to provide an intuitive interpretation of how a cryptographic random scrambling can relax RIP of the measurement matrix and substantially reduce the decoding complexity, i.e., parallel reconstruction. Moreover, some other features owned by a basic CS paradigm, such as robust to packet loss and noise, are also observed.

We now consider a 2D image $\mathbf{X}$ with $M = n \times n$ pixels. If the chosen parameterized transform is RPFrCT, the basis for $\mathbf{X}$ is $(\mathbf{R}_\beta^T \otimes \mathbf{R}_\alpha^T)$ according to Eq. (3.33). Following the same approach adopted in [103], the encoding stage can be written as

$$(\mathbf{Y}) = [\mathbf{y}_1, \mathbf{y}_2, \dots, \mathbf{y}_n]^T = \Phi(\mathbf{X}), \qquad (3.41)$$

where Φ is the product of the $K \times M$ key-dependent sensing matrix $\mathbf{A}_K$ and the $M \times M$ key-dependent basis $\mathbf{\Psi}_K^{-1}$ having the form

$$\mathbf{\Psi}_K^{-1} = \mathbf{D}^{-1} \mathbf{P}^T (\mathbf{R}_\beta^T \otimes \mathbf{R}_\alpha^T), \qquad (3.42)$$

and

$$\mathbf{A}_K = \begin{bmatrix} \mathbf{A}_1 & & & \\ & \mathbf{A}_2 & & \\ & & \ddots & \\ & & & \mathbf{A}_n \end{bmatrix} \qquad (3.43)$$

with $\mathbf{A}_j = \mathbf{A}$ for $j \in \{1, \cdots n\}$ being Gaussian matrices. As we discussed in Sect. 3.4.4.1, repeatedly using the same sensing matrix for different signal segments

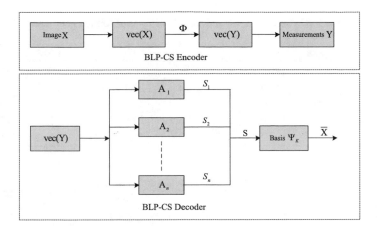

Fig. 3.22 Block diagram of BLP-CS for digital images

can speed up the reconstruction if some off-line mechanism is allowed to calculate the pseudoinverse of $\mathbf{A}$ in advance.

According to Sects. 3.4.2 and 3.4.3, $(\mathbf{S}) = [\mathbf{s}_1, \mathbf{s}_2, \ldots, \mathbf{s}_n]^T = \mathbf{\Psi}_K^{-1}(\mathbf{X})$ is sparse in the canonical basis. A parallel construction is applied as

$$\min \|\mathbf{s}_j\|_1 \ \text{subject to} \ \mathbf{y}_j = \mathbf{A}\mathbf{s}_j. \tag{3.44}$$

for all $j \in \{1, 2, \ldots, n\}$. Finally, the recovered image is given by $(\bar{\mathbf{X}}) = \mathbf{\Psi}_K(\mathbf{S})$. A block diagram of the whole system is depicted in Fig. 3.22. In summary, this system is a instance of the simplified BLP-CS model.

To further illustrate how the random scrambling $\mathbf{P}$ relaxes the RIP requirement of the sensing matrix $\mathbf{A}$, we consider another sampling configuration

$$(\mathbf{Y}) = \mathbf{\Phi}(\mathbf{X}), \tag{3.45}$$

where $\mathbf{\Phi} = \mathbf{A}_K \hat{\mathbf{\Psi}}_K^{-1}$ with $\mathbf{A}_K$ is the same as defined above and $\hat{\mathbf{\Psi}}_K^{-1} = \mathbf{D}^{-1}(\mathbf{R}_\beta^T \otimes \mathbf{R}_\alpha^T)$. Here, we note that the only difference of $\mathbf{\Psi}_K^{-1}$ and $\hat{\mathbf{\Psi}}_K^{-1}$ is the permutation matrix $\mathbf{P}$. The reconstruction is exactly the same as that of Eq. (3.44). By construction, this is a special form of BCS [104], where each block is a column of the frequency coefficients, together with block independent recovery. We call this model BCS-In. We also note that using the smoothed projected Landweber operator can largely improve the BCS reconstrution quality at relatively low extra computation overhead [105]. However, the study of embedding the smoothed projected Landweber operator in the BLP-CS reconstruction is out of the scope of this paper.

Four representative images, "Lena", "Peppers", "Cameraman" and "Baboon" of size 512×512 are used as our test images. The tests are carried out under different

Table 3.2 Comparison between BLP-CS and BCS-In in terms of APSNR at different SRs. (A and B stand for BLP-CS and BCS-In, respectively.)

SR	10%		30%		50%		70%	
Model	A	B	A	B	A	B	A	B
Lena	21.6	15.5	27.5	23.3	31.4	27.3	35.7	32.1
Peppers	20.9	14.4	27.2	22.6	30.9	27.9	34.7	32.5
Cameraman	19.2	13.0	24.8	21.5	28.6	27.4	32.9	32.8
Baboon	17.8	9.7	20.2	17.6	22.6	21.3	25.8	25.2

Table 3.3 APSNR of the reconstructions under AWGN and various PLRs

SR	0.1	0.3	0.5	0.7
Ideal BLP-CS	21.6	27.5	31.4	35.7
BLP-CS AWGN	21.8	27.4	31.3	34.9
BLP-CS 10% PLR	21.7	26.8	30.5	34.1
BLP-CS 20% PLR	20.9	26.2	29.5	32.7
BLP-CS 30% PLR	19.9	25.5	28.5	31.3

sampling rate $SR = \frac{K}{M} \times 100\%$. The reconstruction quality is evaluated in terms of average[4] peak signal-to-noise ratio, $APSNR (dB) = 10 \cdot \log_{10} \mathbb{E}\left(\frac{M255^2}{\|(X)-(\tilde{X})\|_2^2}\right)$. The results are listed in Table 3.2, from which it can be found a cryptographic random scrambling helps make the column sparsity level of **S** uniform. The last point worth mentioning is that random scrambling is suitable for all kind of 2D sparse data (all kind of sparsifying coefficients under parameterized orthonormal transform), which extends the result that zig-zag scrambling works for DCT2 coefficients [50].

The basic CS paradigm that works under RIP theory is known to be robust with respect to transmission imperfections such as noise or packet loss [106, 107]. Since the new proposal works under the RIPless theory at only the encoder but RIP theory at the decoder, we expect the same property in our approach. To quantitatively study this, we evaluate the robustness of the proposed framework with respect to AWGN and various packet loss rates (PLRs). In the former case, we artificially add a zero-mean normal distribution random sequence with variance 1 to the measurements while in the latter we randomly discard certain number of measurements governed by PLR. Then we perform reconstruction on the corrupted measurements. In real applications, PLR can be up to 30% [108] and we measure the quality of the reconstruction in terms of APSNR at 10%, 20% and 30% PLR, respectively. These tests were carried out using the "Lena" image, but similar results were obtained using other images. As observed from Table 3.3, our scheme is almost immune to AWGN when we compare the APSNR of the ideal case and the one with AWGN. In addition, comparing the

---

[4]$\mathbb{E}$ denotes calculate average over 100 tests.

APSNRs at different levels of PLR, we found that the reduction rate of APSNR is linear to the increasing rate of PLR, which implies that all measurements are roughly of the same importance [106].

3.5   Concluding Remarks

In this chapter, we investigated multimedia data encryption based on CS together with some other techniques such as chaos theory, optical transform, image processing, etc. Some possible encryption frameworks were firstly proposed. Then, an encryption scheme for PCS was proposed. It was found from simulation results that the proposed scheme possesses a high robustness against AWGN and cropping attack. Security analysis indicates that the asymptotic spherical secrecy is achievable. Moreover, a scalable encryption framework with involvement of image processing techniques was presented to protect significant image regions. Biometrics and sub-sampling motivated key generation method makes use of CSRM and CRCS theories to generate two keys, serving for the significant block encryption and the insignificant block encryption, respectively. The framework is comprised of two different types of encryption models, including sensitive encryption and robust encryption, corresponding to the significant block encryption and the insignificant block encryption, respectively. It can work efficiently in a parallel computing environment, bring about progressive transmission, and deliver the effect of double protection. Finally, We suggested a BLP-CS model by making use of the non-RIP measurement matrix construction. The security of the BLP-CS model is discussed from various aspects, such as brute-force attack, ciphertext-only attack and plaintext attacks. The proposed model can be used for secure compressive image sampling. Both theoretical analyses and experimental results support our expectation.

References

1. Z. Hua, Y. Zhou, Image encryption using 2D Logistic-adjusted-Sine map. Inf. Sci. **339**, 237–253 (2016)
2. Y. Zhang, D. Xiao, An image encryption scheme based on rotation matrix bit-level permutation and block diffusion. Commun. Nonlinear Sci. Numer. Simu. **19**(1), 74–82 (2014)
3. Z. Hua, Y. Zhou, C.-M. Pun, C.P. Chen, 2D Sine Logistic modulation map for image encryption. Inf. Sci. **297**, 80–94 (2015)
4. Z. Hua, Y. Zhou, Design of image cipher using block-based scrambling and image filtering. Inf. Sci. **396**, 97–113 (2017)
5. Y. Zhang, D. Xiao, Y. Shu, J. Li, A novel image encryption scheme based on a linear hyperbolic chaotic system of partial differential equations. Signal Process. Image Commun. **28**(3), 292–300 (2013)
6. Z. Hua, F. Jin, B. Xu, H. Huang, 2D Logistic-Sine-coupling map for image encryption. Signal Process. **149**, 148–161 (2018)

7. J.-X. Chen, Z.-L. Zhu, C. Fu, L.-B. Zhang, Y. Zhang, An image encryption scheme using non-linear inter-pixel computing and swapping based permutation approach. Commun. Nonlinear Sci. Numer. Simu. **23**(1–3), 294–310 (2015)
8. Z. Hua, S. Yi, Y. Zhou, Medical image encryption using high-speed scrambling and pixel adaptive diffusion. Signal Process. **144**, 134–144 (2018)
9. J.-X. Chen, Z.-L. Zhu, C. Fu, H. Yu, Y. Zhang, Reusing the permutation matrix dynamically for efficient image cryptographic algorithm. Signal Process. **111**, 294–307 (2015)
10. R. Lan, J. He, S. Wang, T. Gu, X. Luo, Integrated chaotic systems for image encryption. Signal Process. **147**, 133–145 (2018)
11. L.Y. Zhang, X. Hu, Y. Liu, K.-W. Wong, J. Gan, A chaotic image encryption scheme owning temp-value feedback. Commun. Nonlinear Sci. Numer. Simu. **19**(10), 3653–3659 (2014)
12. Y. Zhang, D. Xiao, Self-adaptive permutation and combined global diffusion for chaotic color image encryption. AEU-Int. J. Electron. Commun. **68**(4), 361–368 (2014)
13. W. Wen, Y. Zhang, Z. Fang, J.-X. Chen, Infrared target-based selective encryption by chaotic maps. Opt. Commun. **341**, 131–139 (2015)
14. J.-X. Chen, Z.-L. Zhu, Z. Liu, C. Fu, L.-B. Zhang, H. Yu, A novel double-image encryption scheme based on cross-image pixel scrambling in gyrator domains. Opt. Express **22**(6), 7349–7361 (2014)
15. Y. Zhang, D. Xiao, W. Wen, Y. Tian, Edge-based lightweight image encryption using chaos-based reversible hidden transform and multiple-order discrete fractional cosine transform. Opt. Laser Tech. **54**, 1–6 (2013)
16. J.-X. Chen, Z.-L. Zhu, C. Fu, L.-B. Zhang, Y. Zhang, Cryptanalysis and improvement of an optical image encryption scheme using a chaotic Baker map and double random phase encoding. J. Opt. **16**(12), 125403 (2014)
17. J.-X. Chen, Z.-L. Zhu, C. Fu, L.-B. Zhang, H. Yu, Analysis and improvement of a double-image encryption scheme using pixel scrambling technique in gyrator domains. Opt. Lasers Eng. **66**, 1–9 (2015)
18. J.-X. Chen, Z.-L. Zhu, C. Fu, H. Yu, Optical image encryption scheme using 3-D chaotic map based joint image scrambling and random encoding in gyrator domains. Opt. Commun. **341**, 263–270 (2015)
19. S. Liansheng, Z. Bei, N. Xiaojuan, T. Ailing, Optical multiple-image encryption based on the chaotic structured phase masks under the illumination of a vortex beam in the gyrator domain. Opt. Express **24**(1), 499–515 (2016)
20. N. Zhou, A. Zhang, F. Zheng, L. Gong, Novel image compression-encryption hybrid algorithm based on key-controlled measurement matrix in compressive sensing. Opt. Laser Techn. **62**, 152–160 (2014)
21. N. Zhou, A. Zhang, J. Wu, D. Pei, Y. Yang, Novel hybrid image compression-encryption algorithm based on compressive sensing. Optik **125**(18), 5075–5080 (2014)
22. S.N. George, D.P. Pattathil, A novel approach for secure compressive sensing of images using multiple chaotic maps. J. Opt. **43**(1), 1–17 (2014)
23. J. Lang, J. Zhang, Optical image cryptosystem using chaotic phase-amplitude masks encoding and least-data-driven decryption by compressive sensing. Opt. Commun. **338**, 45–53 (2015)
24. H. Liu, D. Xiao, Y. Liu, Y. Zhang, Securely compressive sensing using double random phase encoding. Optik **126**(20), 2663–2670 (2015)
25. Y. Zhang, L.Y. Zhang, Exploiting random convolution and random subsampling for image encryption and compression. Electron. Lett. **51**(20), 1572–1574 (2015)
26. L. Y. Zhang, K.-W. Wong, Y. Zhang, Q. Lin, Joint quantization and diffusion for compressed sensing measurements of natural images, in *Proceedings of IEEE International Symposium on Circuits and System, ISCAS* (2015), pp. 2744–2747
27. L.Y. Zhang, K.-W. Wong, Y. Zhang, J. Zhou, Bi-level protected compressive sampling. IEEE Trans. Multimed. **18**(9), 1720–1732 (2016)
28. J. Li, J.S. Li, Y.Y. Pan, R. Li, Compressive optical image encryption. Sci. Rep. **5**, 10374 (2015)
29. Y. Zhang, J. Zhou, F. Chen, L.Y. Zhang, K.-W. Wong, H. Xing, D. Xiao, Embedding cryptographic features in compressive sensing. Neurocomputing **205**, 472–480 (2016)

30. R. Fay, Introducing the counter mode of operation to compressed sensing based encryption. Inf. Process. Lett. **116**(4), 279–283 (2016)
31. Y. Zhang, J. Zhou, F. Chen, L.Y. Zhang, D. Xiao, B. Chen, L. Xiaofeng, A block compressive sensing based scalable encryption framework for protecting significant image regions. Int. J. Bifurcat. Chaos **26**(11), 1650191 (2016)
32. H. Huang, X. He, Y. Xiang, W. Wen, Y. Zhang, A compression-diffusion-permutation strategy for securing image. Signal Process. **150**, 183–190 (2018)
33. D. Zhang, X. Liao, B. Yang, Y. Zhang, A fast and efficient approach to color-image encryption based on compressive sensing and fractional Fourier transform. Multimed. Tools Appl. **77**(2), 2191–2208 (2018)
34. J. Chen, Y. Zhang, L.Y. Zhang, On the security of optical ciphers under the architecture of compressed sensing combining with double random phase encoding. IEEE Photonics J. **9**(4), 1–11 (2017)
35. X. Chai, Z. Gan, Y. Chen, Y. Zhang, A visually secure image encryption scheme based on compressive sensing. Signal Process. **134**, 35–51 (2017)
36. N. Zhou, J. Yang, C. Tan, S. Pan, Z. Zhou, Double-image encryption scheme combining DWT-based compressive sensing with discrete fractional random transform. Opt. Commun. **354**, 112–121 (2015)
37. X. Chai, X. Zheng, Z. Gan, D. Han, Y. Chen, An image encryption algorithm based on chaotic system and compressive sensing. Signal Process. **148**, 124–144 (2018)
38. Y. Zhang, L.Y. Zhang, J. Zhou, L. Liu, F. Chen, X. He, A review of compressive sensing in information security field. IEEE Access **4**, 2507–2519 (2016)
39. X. Li, X. Meng, X. Yang, Y. Yin, Y. Wang, X. Peng, W. He, G. Dong, H. Chen, Multiple-image encryption based on compressive ghost imaging and coordinate sampling. IEEE Photonics J. **8**(4), 1–11 (2016)
40. G. Hu, D. Xiao, Y. Wang, T. Xiang, An image coding scheme using parallel compressive sensing for simultaneous compression-encryption applications. J. Visual Commun. Image Represent. **44**, 116–127 (2017)
41. G. Hu, D. Xiao, Y. Wang, T. Xiang, Q. Zhou, Securing image information using double random phase encoding and parallel compressive sensing with updated sampling processes. Opt. Lasers Eng. **98**, 123–133 (2017)
42. E.J. Candès, T. Tao, Near-optimal signal recovery from random projections: Universal encoding strategies? IEEE Trans. Inf. Theory **52**(12), 5406–5425 (2006)
43. R. Huang, K. Rhee, S. Uchida, A parallel image encryption method based on compressive sensing. Multimed. Tools Appl. **72**(1), 71–93 (2014)
44. Y. Rachlin, D. Baron, The secrecy of compressed sensing measurements, in *Proceedings of 46th Annual Allerton Conference on Communication, Control and Computing*, Urbana-Champaign, IL (2008), pp. 813–817
45. A. Orsdemir, H.O. Altun, G. Sharma, M.F. Bocko, On the security and robustness of encryption via compressed sensing, in *Proceedings of IEEE Military Communications Conference (MILCOM)*, San Diego, CA (2008), pp. 1–7
46. S.A. Hossein, A. Tabatabaei, N. Zivic, Security analysis of the joint encryption and compressed sensing, in *Proceedings of 20th Telecommunications Forum (TELFOR)*, Belgrade (2012), pp. 799–802
47. T. Bianchi, V. Bioglio, E. Magli, On the security of random linear measurements, in *Proceedings of IEEE International Conference on Acoustics, Speech, and Signal Processing (ICASSP)*, Florence (2014), pp. 3992–3996
48. S.N. George, D.P. Pattathil, A secure LFSR based random measurement matrix for compressive sensing. Sens. Imag. **15**(1), 1–29 (2014)
49. L. Yu, J.P. Barbot, G. Zheng, H. Sun, Compressive sensing with chaotic sequence. IEEE Signal Process. Lett. **17**(8), 731–734 (2010)
50. H. Fang, S.A. Vorobyov, H. Jiang, O. Taheri, Permutation meets parallel compressed sensing: How to relax restricted isometry property for 2D sparse signals. IEEE Trans. Signal Process. **62**(1), 196–210 (2014)

51. P. Refregier, B. Javidi, Optical image encryption using input plane and Fourier plane random encoding, in *SPIE, International Society for Optics and Photonics* (1995), pp. 62–68
52. P. Lu, Z. Xu, X. Lu, X. Liu, Digital image information encryption based on compressive sensing and double random-phase encoding technique. Optik **124**(16), 2514–2518 (2013)
53. N. Rawat, R. Kumar, B.-G. Lee, Implementing compressive fractional Fourier transformation with iterative kernel steering regression in double random phase encoding. Optik **125**(18), 5414–5417 (2014)
54. H. Takeda, S. Farsiu, P. Milanfar, Kernel regression for image processing and reconstruction. IEEE Trans. Image Process. **16**(2), 349–366 (2007)
55. B. Deepan, C. Quan, Y. Wang, C. Tay, Multiple-image encryption by space multiplexing based on compressive sensing and the double-random phase-encoding technique. Appl. Opt. **53**(20), 4539–4547 (2014)
56. Q. Gong, X. Liu, G. Li, Y. Qin, Multiple-image encryption and authentication with sparse representation by space multiplexing. Appl. Opt. **52**(31), 7486–7493 (2013)
57. U. Gopinathan, D.S. Monaghan, T.J. Naughton, J.T. Sheridan, A known-plaintext heuristic attack on the Fourier plane encryption algorithm. Opt. Express **14**(8), 3181–3186 (2006)
58. X. Peng, P. Zhang, H. Wei, B. Yu, Known-plaintext attack on optical encryption based on double random phase keys. Opt. Lett. **31**(8), 1044–1046 (2006)
59. Y. Frauel, A. Castro, T.J. Naughton, B. Javidi, Resistance of the double random phase encryption against various attacks. Opt. Express **15**(16), 10253–10265 (2007)
60. Y. Zhang, D. Xiao, W. Wen, H. Liu, Vulnerability to chosen-plaintext attack of a general optical encryption model with the architecture of scrambling-then-double random phase encoding. Opt. Lett. **38**(21), 4506–4509 (2013)
61. Y. Rivenson, A. Stern, B. Javidi, Single exposure super-resolution compressive imaging by double phase encoding. Opt. Express **18**(14), 15094–15103 (2010)
62. A. Alfalou, C. Brosseau, Optical image compression and encryption methods. Adv. Opt. Photonic. **1**(3), 589–636 (2009)
63. A. Alfalou, C. Brosseau, Exploiting root-mean-square time-frequency structure for multiple-image optical compression and encryption. Opt. Lett. **35**(11), 1914–1916 (2010)
64. A. Alfalou, C. Brosseau, N. Abdallah, M. Jridi, Simultaneous fusion, compression, and encryption of multiple images. Opt. Express **19**(24), 24023–24029 (2011)
65. W.-K. Yu, M.-F. Li, X.-R. Yao, X.-F. Liu, L.-A. Wu, G.-J. Zhai, Adaptive compressive ghost imaging based on wavelet trees and sparse representation. Opt. Express **22**(6), 7133–7144 (2014)
66. G. Oliveri, L. Poli, P. Rocca, A. Massa, Bayesian compressive optical imaging within the Rytov approximation. Opt. Lett. **37**(10), 1760–1762 (2012)
67. J. Greenberg, K. Krishnamurthy, D. Brady, Compressive single-pixel snapshot X-ray diffraction imaging. Opt. Lett. **39**(1), 111–114 (2014)
68. X. Lin, G. Wetzstein, Y. Liu, Q. Dai, Dual-coded compressive hyperspectral imaging. Opt. Lett. **39**(7), 2044–2047 (2014)
69. X. Liu, Y. Cao, P. Lu, X. Lu, Y. Li, Optical image encryption technique based on compressed sensing and arnold transformation. Optik **124**(24), 6590–6593 (2013)
70. Y. Zhang, D. Xiao, Double optical image encryption using discrete Chirikov standard map and chaos-based fractional random transform. Opt. Lasers Eng. **51**(4), 472–480 (2013)
71. X. Liu, W. Mei, H. Du, Optical image encryption based on compressive sensing and chaos in the fractional Fourier domain. J. Modern Opt. **61**(19), 1570–1577 (2014)
72. W.-X. Wang, R. Yang, Y.-C. Lai, V. Kovanis, C. Grebogi, Predicting catastrophes in nonlinear dynamical systems by compressive sensing. Phys. Rev. Lett. **106**(15), 154101 (2011)
73. W.-X. Wang, R. Yang, Y.-C. Lai, V. Kovanis, M.A.F. Harrison, Time-series-based prediction of complex oscillator networks via compressive sensing. Eur. Phys. Lett. **94**(4), 48006 (2011)
74. V. Cambareri, M. Mangia, F. Pareschi, R. Rovatti, G. Setti, Low-complexity multiclass encryption by compressed sensing. IEEE Trans. Signal Process. **63**(9), 2183–2195 (2015)
75. M.-R. Zhang, G.-C. Shao, K.-C. Yi, T-matrix and its applications in image processing. Electron. Lett. **40**(25), 1583–1584 (2004)

76. X. Wang, L. Teng, X. Qin, A novel colour image encryption algorithm based on chaos. Signal Process. **92**(4), 1101–1108 (2012)
77. M. Frunzete, L. Yu, J. Barbot, A. Vlad, Compressive sensing matrix designed by tent map, for secure data transmission, in *Proceedings of IEEE Signal Process: Algorithms Architectures Arrangements and Applications (SPA)*, Poznan (2011), pp. 1–6
78. M. Grant, S. Boyd, Y. Ye, CVX: Matlab software for disciplined convex programming (2008)
79. Wikipedia (2018), https://en.wikipedia.org/wiki/Sobel_operator
80. Y. Zhou, Z. Hua, C.-M. Pun, C.P. Chen, Cascade chaotic system with applications. IEEE Trans. Cyber. **45**(9), 2001–2012 (2015)
81. T.T. Do, L. Gan, N.H. Nguyen, T.D. Tran, Fast and efficient compressive sensing using structurally random matrices. IEEE Trans. Signal Process. **60**(1), 139–154 (2012)
82. C.E. Shannon, Communication theory of secrecy systems. Bell Syst. Tech. J. **28**(4), 656–715 (1949)
83. J. Romberg, Compressive sensing by random convolution. SIAM J. Imag. Sci. **2**(4), 1098–1128 (2009)
84. G. Bhatnagar, Q.J. Wu, Biometric inspired multimedia encryption based on dual parameter fractional Fourier transform. IEEE Trans. Syst. Man Cybern. Syst. **44**(9), 1234–1247 (2014)
85. F. Hao, R. Anderson, J. Daugman, Combining crypto with biometrics effectively. IEEE Trans. Comput. **55**(9), 1081–1088 (2006)
86. S.V. Gaddam, M. Lal, Efficient cancelable biometric key generation scheme for cryptography. Int. J. Netw Security **11**(2), 61–69 (2010)
87. J.-X. Chen, Z.-L. Zhu, C. Fu, L.-B. Zhang, Y. Zhang, An efficient image encryption scheme using lookup table-based confusion and diffusion. Nonlinear Dyn., 1–16 (2015)
88. M.A. Figueiredo, R.D. Nowak, S.J. Wright, Gradient projection for sparse reconstruction: application to compressed sensing and other inverse problems. IEEE J. Sel. Top. Signal Process. **1**(4), 586–597 (2007)
89. B. Zeng, S.-K.A. Yeung, S. Zhu, M. Gabbouj, Perceptual encryption of H. 264 videos: embedding sign-flips into the integer-based transforms. IEEE Trans. Inf. Foren. Sec. **9**(2), 309–320 (2014)
90. A. Pande, J. Zambreno, The secure wavelet transform. J. Real-Time Image Process. **7**(2), 131–142 (2012)
91. A. Pande, P. Mohapatra, J. Zambreno, Securing multimedia content using joint compression and encryption. IEEE Multimed. **20**(4), 50–61 (2013)
92. D. Engel, A. Uhl, Parameterized biorthogonal wavelet lifting for lightweight JPEG 2000 transparent encryption, in *Proceedings of 7th Workshop Multimedia and Security* (2005), pp. 63–70
93. S.-K.A. Yeung, B. Zeng, A new design of multiple transforms for perceptual video encryption, in *Proceedings of 19th IEEE International Conference on Image Processing, ICIP* (2012), pp. 2637–2640
94. G. Unnikrishnan, J. Joseph, K. Singh, Optical encryption by double-random phase encoding in the fractional Fourier domain. Opt. Lett. **25**(12), 887–889 (2000)
95. P. Refregier, B. Javidi, Optical image encryption based on input plane and Fourier plane random encoding. Opt. Lett. **20**(7), 767–769 (1995)
96. I. Venturini, P. Duhamel, Reality preserving fractional transforms, in *Proceedings of IEEE International Conference on Acoustics, Speech and Signal Processing, ICASSP* (2004), pp. 205–208
97. G. Cariolaro, T. Ersehe, P. Kraniaukas, The fractional discrete cosine transform. IEEE Trans. Signal Process. **50**(4), 902–911 (2002)
98. J. Tropp, A.C. Gilbert, Signal recovery from random measurements via orthogonal matching pursuit. IEEE Trans. Inf. Theory **53**(12), 4655–4666 (2007)
99. S. Mun, J.E. Fowler, DPCM for quantized block-based compressed sensing of images, in *Proceedings of European Signal Processing Conference* (2012), pp. 1424–1428
100. L. Tong, F. Dai, Y. Zhang, J. Li, D. Zhang, Compressive sensing based video scrambling for privacy protection, in *Proceedings of IEEE Visual Communications and Image Processing, VCIP* (2011), pp. 1–4

101. V. Cambareri, M. Mangia, F. Pareschi, R. Rovatti, G. Setti, On known-plaintext attacks to a compressed sensing-based encryption: a quantitative analysis. IEEE Trans. Inf. Foren. Sec. **10**(10), 2182–2195 (2015)
102. E.J. Candès, T. Tao, Decoding by linear programming. IEEE Trans. Inf. Theory **51**(12), 4203–4215 (2005)
103. M.F. Duarte, M.A. Davenport, D. Takhar, J.N. Laska, T. Sun, K.E. Kelly, R.G. Baraniuk, Single-pixel imaging via compressive sampling. IEEE Signal Process. Mag. **25**(2), 83 (2008)
104. L. Gan, Block compressed sensing of natural images, in *Proceedings of 15th International Conference on Digital Signal Processing* (2007), pp. 403–406
105. J.E. Fowler, S. Mun, E.W. Tramel, Multiscale block compressed sensing with smoothed projected landweber reconstruction, in *Proceedings of 19th European Signal Processing Conference* (IEEE, 2011), pp. 564–568
106. J.N. Laska, P.T. Boufounos, M.A. Davenport, R.G. Baraniuk, Democracy in action: quantization, saturation, and compressive sensing. Appl. Comput. Harmon. Anal. **31**(3), 429–443 (2011)
107. A.G. Dimakis, P.O. Vontobel, LP decoding meets LP decoding: a connection between channel coding and compressed sensing, in *Proceedings of 47th Annual Allerton Conference on Communication, Control, and Computing* (IEEE, 2009), pp. 8–15
108. J. Zhao, R. Govindan, Understanding packet delivery performance in dense wireless sensor networks, in *Proceedings of 1st International Conference on Embedded Networked Sensor Systems* (2003), pp. 1–13

Chapter 4
Cloud Computing Security

Abstract The reconstruction of CS is a time-consuming process, which is unfavorable for resources-limited terminals in application. Cloud computing resources are abundant and can be used for being responsible for the reconstruction. However, privacy-preserved outsourcing mechanism needs to be elaborately decorated since the cloud is often not fully trusted. In this chapter, we design two CS reconstruction outsourcing mechanisms in multi-clouds, which are sparse reconstruction service and sparse robustness decoding service, respectively. It is shown that these two mechanisms are secure, efficient, and feasible.

4.1 Compressive Sensing Meets Cloud Computing

The sampling complexity of CS is linear while the reconstruction complexity is cubic. For resource-constrained devices, to reconstruct and store signals is unpractical because it would consume too much computational resources and occupy too much storage space. Clouds can be a promising platform for this thanks to their powerful computation and storage capabilities. However, the untrust nature of cloud environments makes privacy-assured CS reconstruction outsourcing essential. Computation outsourcing has been studied for many years and the computation tasks harnessed by the resource-constrained clients can be off-loaded to powerful computation devices like cloud servers. The input and output privacy of outsourcing computation to the clouds needs to be protected as these tasks are often sensitive and the clouds might not be honest to the clients. Thus, if consideration must be given to solving the heavy reconstruction work while guaranteeing the privacy of the signal, one of the best choices is to securely outsource it to a cloud with abundant computing resources.

63
Y. Zhang et al., *Secure Compressive Sensing in Multimedia Data,*
Cloud Computing and IoT, SpringerBriefs in Signal Processing,
https://doi.org/10.1007/978-981-13-2523-6_4

4.2   Privacy-Preserving Compressive Sensing Reconstruction in Public Cloud

Consider a scenario that a limited-resource owner shares the signal data of interest with users by resorting to the cloud. The owner captures the CS measurements by sampling the signal of interest for the purpose of storage overhead reduction and then sends them to the cloud for storage and sparse reconstruction purpose. When the users need the signal, the cloud performs the sparse reconstruction service for the users. However, the owner hopes that the cloud cannot observe the signal but the users because the cloud runs in an open environment operated by external third parties. In this scenario, the owner requires to harness the cloud for secure outsourcing of sparse reconstruction service for CS.

We propose to explore such a privacy-assured parallel outsourcing of sparse reconstruction service (POSRS). We focus on the privacy of support-set for a sparse signal, since the positions of the nonzero entries can be fully determined by support-set. Moreover, the support-set-assured mechanism is compatible with the storage characteristic of the sparse signal that only requires recording indices and magnitudes of non-zero entries. Assume that multi-clouds are semi-trusted and cannot collude with each other in private. In recent years, dealing with multi-clouds in cloud computing security is becoming more popular than a single cloud because of service availability failure and the possibility that there are malicious insiders in the single cloud [1]. The signal owner exploits a simple exchange primitive to encrypt the support-set. This primitive can achieve a similar scrambling effect as that of random permutation matrix that the RIP for 2D sparse signals with high probability can be relaxed [2]. However, it consumes lower complexity and less memory space in comparison with the latter. After encryption, the owner performs parallel CS for the signal column by column with the same sensing matrix. Then, the compressive measurements and support-set are distributed over multi-clouds for storage and reconstruction service. Each cloud only has a small amount of information of both the measurements and support-set. After the reconstruction, each cloud only obtains the encrypted entries and has partial index information asymmetric to these entries in hand so that the security can be guaranteed. When the user requires the signal, the multi-clouds reconstruct respective signal fragments in parallel. The user integrates these fragments into a signal and decrypts it to form the signal of interest.

4.2.1   Design of Privacy-Assured POSRS in Multi-clouds

4.2.1.1   Service and Threat Model

Figure 4.1 depicts the service model for the POSRS architecture in public multi-clouds. The basic message flow is as follows. The owner firstly encrypts the signal with a key known both the owner and the users and performs PCS to reduce storage

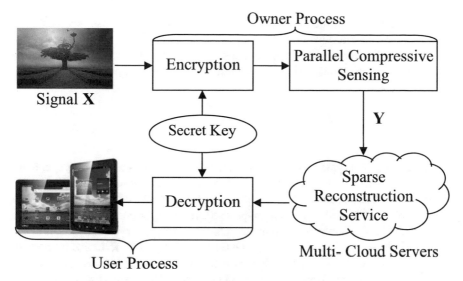

Fig. 4.1 The POSRS architecture

space. Then the generated measurements are packeted and outsourced to the multi-clouds server for storage and processing. Once upon receiving the requests from the users, the multi-clouds will on-demand fulfill the sparse reconstruction service in parallel.

Our model is a little akin to that in [3]. The similarities lie in that the users are assumed to possess only limited computational resources and that the threat comes from a semi-trusted cloud as the adversary, i.e., the cloud is assumed to honestly exploit the sparse reconstruction service as specified, but be curious in learning the owner/users' data content. However, an obvious difference is that encryption is behind CS in the latter while in our model, the encryption is prior to CS. In addition, we consider the case of multi-clouds and parallel outsourcing. At the first glance, the encryption followed by compressive sampling seems irrational due to the fact that the encryption must break the correlation between data while the classic multimedia compression schemes like JPEG 2000 always benefit from strong correlation. Interestingly, the proposed encryption technique cannot affect but enhance the compression performance of parallel CS [2], which is mainstay in our model. Note that we assume that the multi-clouds cannot collude with each other in private.

4.2.1.2 Design Goals

The proposed POSRS should achieve the following three goals.

- Security. The service model can assure the privacy of the owner's sparse data, of which the cloud cannot get meaning knowledge.

- Effectiveness. The cloud is allowed to effectively exploit the sparse reconstruction service over the encrypted data.
- Efficiency. The computation and storage savings can be obtained on the owner and user sides while the encryption should not result in the computation augment on the cloud side.

4.2.1.3 Basic Framework

Compressing an encrypted signal, as previously mentioned, is a big challenge, however, it is allowed to exploit the parallel CS for a permutation-encryption based 2D sparse signal. Let us start with a 2D sparse signal $\mathbf{X}$. A signal owner first gets the support set $\mathbf{s}_v$ of $\mathbf{X}$ and shares an adaptive secret encryption operator $\mathbf{p}$ with the users, which is not known by the multi-clouds. The owner encrypts the indices of the non-zero entries in $\mathbf{X}$ with $\mathbf{p}$ to obtain the encrypted signal $\mathbf{X}'$ followed by parallel CS with the same sensing matrix $\mathbf{\Phi}$ known to both the owner and the clouds. The obtained measurements $\mathbf{Y}$ and support set $\mathbf{s}_v$ are grouped and sent to the multi-clouds for storage and sparse reconstruction processing. When the multi-clouds fulfill the sparse reconstruction service, the reconstructed signal data $\mathbf{X}''$ is still in encrypted form so that the signal privacy is preserved. Each cloud cannot immediately observes the original signal $\mathbf{X}$ with a small amount of $\mathbf{p}$ and $\mathbf{Y}$. Note that $|\cdot|$ represents the corresponding size. The design of POSRS is a collection of six polynomial-time algorithms marked as $\Omega = (SupGet, KeyGen, SigEnc, ParCS, SRSCS, SigDec)$.

- $SupGet\,(\mathbf{X}) \rightarrow \mathbf{s}_v$ is a support set vector generation algorithm that is run by the signal owner. It takes a signal $\mathbf{X}$ as input and returns a support vector $\mathbf{s}_v$.
- $KeyGen\,(1^p) \rightarrow \mathbf{p}$ is a key generation algorithm that is run by the signal owner. It takes a security parameter p as input and returns an encryption operator $\mathbf{p}$ with $|\mathbf{p}| = |\mathbf{s}_v|$.
- $SigEnc\,(\mathbf{X}, \mathbf{p}) \rightarrow \mathbf{X}'$ is a signal encryption algorithm performed by the signal owner who encrypts a signal $\mathbf{X}$ with $\mathbf{p}$ and returns the encrypted signal $\mathbf{X}'$.
- $ParCS\,(\mathbf{X}', \mathbf{\Phi}) \rightarrow \mathbf{Y}$ is a parallel CS algorithm running at the signal owner side. It takes as inputs an encrypted signal $\mathbf{X}'$ and a sensing matrix $\mathbf{\Phi}$, and returns the measurements $\mathbf{Y}$.
- $SRSCS\,(\mathbf{Y}, \mathbf{\Phi}) \rightarrow \mathbf{X}''$ is an algorithm running at the cloud side to provide the sparse reconstruction service for CS. It takes as inputs the measurements $\mathbf{Y}$ and the sensing matrix $\mathbf{\Phi}$, and returns the reconstructed signal $\mathbf{X}''$.
- $SigDec\,(\mathbf{X}'', P^{-1}) \rightarrow \tilde{\mathbf{X}}$ is a signal decryption algorithm that is handled by the users who decrypts the signal $\mathbf{X}''$ with the encryption operator $\mathbf{p}$, and returns the original signal $\tilde{\mathbf{X}}$.

4.2.1.4 Design Details

With respect to the basic framework of POSRS, the design details will be illustrated in the following.

SupGet. The signal owner changes $\mathbf{X}$ into 1D signal $\mathbf{x}$ of length $n = M \times N$ according to the column mode from left to right and then gets the support vector $\mathbf{s}_v$ of $\mathbf{x}$ with $|\mathbf{s}_v| = s$ because of $\|\mathbf{x}\|_0 = s$, as shown in Algorithm 1. Each entry of $\mathbf{s}_v$ lies in the integer interval $[1, n]$.

Algorithm 8 Support Vector Generation.

Input: $\mathbf{X}$.
Output: $\mathbf{s}_v$.
1: $\mathbf{x} \leftarrow \mathbf{X}$
2: **for** $i = 1$ to n;
3:    **if** $\mathbf{x}(i) \neq 0$;
4:       Write i to $\mathbf{s}_v$;
5:    **end if**
6: **end for**

KeyGen. The owner employs a random number generator as the security parameter p, which is responsible for generating $\mathbf{p}$, as instantiated in Algorithm 2.

Algorithm 9 Key Generation.

Input: p and $U = \{1, 2, \cdots, n\}$.
Output: $\mathbf{p}$.
1: Set $\mathbf{p} = \mathbf{0}_{s \times 1}$;
2: **for** $i = 1$ to s;
3:    Select a random integer j from U;
4:    Set $\mathbf{p}(i) = j$;
5:    Update U by $U \leftarrow U/\{j\}$;
6: **end for**

SigEnc. The $\mathbf{p}$ exerts the encryption on $\mathbf{x}$ and the encrypted result is $\mathbf{x}'$, as shown in Algorithm 3. The encryption indicates the non-zero entries are randomly rearranged. This is equivalent to a random permutation of all the entries.

Algorithm 10 Signal Encryption.

Input: $\mathbf{p}$, $\mathbf{s}_v$ and $\mathbf{x}$.
Output: $\mathbf{x}' = \mathbf{x}$.
1: **for** $i = 1$ to s;
2:    Exchange the two entries $\mathbf{x}(\mathbf{p}(i))$ and $\mathbf{x}(\mathbf{s}_v(i))$ in $\mathbf{x}$;
3: **end for**

ParCS. Prior to sampling, the encrypted data $\mathbf{x}'$ will be partitioned into multiple column data with equal length followed by repetitive sampling per column with the same sensing matrix $\boldsymbol{\Phi}$ which is Gaussian matrix [4], Bernoulli matrix [5] or other matrices known to the cloud servers. Without loss of generality, let the column length be M, i.e, the partition result can be expressed as a matrix $\mathbf{X}'$ of $M \times N$. The parallel CS process returns a $K \times N$ matrix $\mathbf{Y}$, as shown in Algorithm 4.

Algorithm 11 Parallel Compressive Sensing.

Input: $\boldsymbol{\Phi}$ and $\mathbf{X}'$.
Output: $\mathbf{Y}$.
1: Calculate $\mathbf{Y} = \boldsymbol{\Phi}\mathbf{X}'$;

SRSCS. The multi-cloud servers fulfill the sparse reconstruction service by solving a convex optimization problem, as illustrated in Algorithm 5. Likewise, the parallel CS reconstruction architecture has lower complexity than the traditional CS case has. The former complexity is $\mathcal{O}\left(NM^3\right) = \mathcal{O}\left(nM^2\right)$ whereas the latter is $\mathcal{O}\left(n^3\right)$.

Algorithm 12 Sparse Reconstruction Service.

Input: $\boldsymbol{\Phi}$ and $\mathbf{Y}$.
Output: $\mathbf{X}''$.
1: Solve the following convex optimization problem
    $\min \left\|\mathbf{X}''[j]\right\|_1$ $s.t.$ $\mathbf{Y}[j] = \boldsymbol{\Phi}\mathbf{X}''[j]$, $j \in [1, N]$

SigDec. Upon receiving the reconstructed signal $\mathbf{X}''$, the users transform $\mathbf{X}''$ into a 1D form $\mathbf{x}''$ and carry out the decryption using Algorithm 6. The result is $\tilde{\mathbf{x}}$, which is further converted back in to 2D form $\tilde{\mathbf{X}}$. The $\mathbf{s}_v$ can be gathered from multi-cloud servers.

Algorithm 13 Signal Decryption.

Input: $\mathbf{p}$, $\mathbf{s}_v$ and $\mathbf{x}''$.
Output: $\tilde{\mathbf{x}} = \mathbf{x}''$.
1: **for** $i = s$ down to 1;
2:     Exchange the two entries $\mathbf{x}''(\mathbf{p}(i))$ and $\mathbf{x}''(\mathbf{s}_v(i))$ in $\mathbf{x}''$;
3: **end for**

4.2.2   Multi-clouds Scheduling and Security Analysis

The $\mathbf{Y}$ can be distributed over multiple clouds and each cloud side stores a few columns of $\mathbf{Y}$ with the extreme case of one-to-one correspondence. Such a distribution has the following advantages: (a) Security assurance. Each cloud side can be aware of partial information of $\mathbf{X}'$ and cannot infer full information with the knowledge of a limited number of $\mathbf{Y}[j]$, $1 \leq j \leq N$. (b) Parallel processing. The reconstruction service is realized in parallel and makes full use of cloud resources. Moreover, the complexity induced by parallel CS is $O(Kn)$. If a single sensing matrix is directly to sample $\mathbf{x}'$ and then acquire the same number of measurements, its size will be $KN \times n$, which occupies so much memory space and the corresponding complexity rises to $\mathcal{O}(KNn)$. It goes to show that parallel CS outperforms the traditional CS during the sampling side in terms of memory and complexity.

In addition, the support vector $\mathbf{s}_v$ needs to be cascaded with $\mathbf{Y}$ and distributed on the multi-clouds for storage and further decryption purpose. For clarity, we give a mathematical illustration. Assume that N columns, each of which has K elements, are allotted to ℓ clouds as averagely as possible, and then the first $\ell - 1$ clouds each have $\lfloor N/\ell \rfloor$ columns while the last one has the remaining $N - (\ell - 1)\lfloor N/\ell \rfloor$ columns. Executing the similar operation for $\mathbf{s}_v$, one can assign $\lfloor s/\ell \rfloor$ entries to each of the first $\ell - 1$ clouds and $s - (\ell - 1)\lfloor s/\ell \rfloor$ entries to the last one. Accordingly, for the i-th ($1 \leq i \leq \ell - 1$) cloud, the receiving data are $\{\mathbf{Y}[(i-1)\lfloor N/\ell \rfloor + j_1] | 1 \leq j_1 \leq \lfloor N/\ell \rfloor\} \| \{\mathbf{s}_v[j_2] | j_2 \in [(i-1)\lfloor s/\ell \rfloor + 1,$ $i\lfloor s/\ell \rfloor]\}$. Apparently, No explicit relations exist between the generated $\{\mathbf{X}'[(i-1)$ $\lfloor N/\ell \rfloor + j_1] | 1 \leq j_1 \leq \lfloor N/\ell \rfloor\}$ after decrypting $\{\mathbf{Y}[(i-1)\lfloor N/\ell \rfloor + j_1] | 1 \leq j_1 \leq$ $\lfloor N/\ell \rfloor\}$ and $\{\mathbf{s}_v[j_2] | j_2 \in [(i-1)\lfloor s/\ell \rfloor + 1, i\lfloor s/\ell \rfloor]\}$, therefore the i-th cloud is not able to infer the useful information with the knowledge of $\{\mathbf{s}_v[j_2] | j_2 \in [(i-1)$ $\lfloor s/\ell \rfloor + 1, i\lfloor s/\ell \rfloor]\}$. On the whole, each cloud only has a small amount of information of both the measurements and asymmetric support-set; therefore, the security can be guaranteed.

Now that the sensing matrix keeps consistent between the owner and the cloud, the owner is also able to transfer the sampling process to the multi-cloud servers at the cost of communication due to transmitting the encrypted signal $\mathbf{X}'$ with larger size than doing $\mathbf{Y}$. This transfer will form a particular multi-cloud outsourcing service, referred to as *compressive cloud sensing*, since the owner straightforwardly transmits the encrypted signal $\mathbf{X}'$ and the support vector $\mathbf{s}_v$ to the multi-clouds in group, and the multi-clouds operate PCS for storage decrease and parallel reconstruction service.

4.2.3 Efficiency Analysis

We investigate the complexity of the proposed framework $\Omega = (SupGet, KeyGen, SigEnc, ParCS, SRSCS, SigDec)$, which is $(\mathcal{O}(n), \mathcal{O}(s), \mathcal{O}(s), \mathcal{O}(Kn),$ $\mathcal{O}(nM^2), \mathcal{O}(s))$, respectively. For the traditional CS reconstruction service, the complexity is $\mathcal{O}(n^3)$. In our POSRS, the signal owner only has the complexity of $\mathcal{O}(Kn + n + 2s)$, where $s \ll n$. Meanwhile, the multi-clouds only have the complexity of $\mathcal{O}(nM^2)$, which is much less than $\mathcal{O}(n^3)$. The user only has the complexity of $\mathcal{O}(s)$. That is to say, the signal owner can reduce its original $\mathcal{O}(n^3)$ work to $\mathcal{O}(nM^2)$ by outsourcing the traditional CS reconstruction service to the multi-clouds in parallel and at the same time, the owner and the users has much low complexity. It is demonstrated that the POSRS gains substantial computation savings and has high efficiency.

4.3 Secure Sparse Robustness Decoding Service in Multi-clouds

We consider the situation where a client needs to transmit 2D sparse signals to some semi-trusted cloud servers over lossy channels through packet-loss networks. The clouds provide storage and decoding service, called sparse robustness decoding service (SRDS), for the user. For efficiency and security purpose, the idea of parallelization is integrated into the encoding and decoding phases, and multiple cloud servers are employed to offer SRDS. Specifically, prior to encoding, a 2D sparse signal is encrypted by changing the indices and amplitudes of its non-zero entries. The encryption algorithm has very low complexity as the non-zero entries make up only a small portion of the sparse signal. Then, the client utilizes PCS technique to encode the encrypted signal for robust coding, where a Gaussian measurement matrix can used for sensing the encrypted signal. A feature of CS is to endow the reconstruction with a high cubic computational complexity while retaining linear sampling complexity, which is beneficial to a resource-constrained client. The corresponding 2D compressive measurements as the encoded data are then sent to the multi-clouds for storage and SRDS, where the measurements are allotted to the multi-clouds as equally as possible. To facilitate the distribution of the compressive measurements to the clouds, each column in compressive measurements is taken as a packet and a certain number of packets constitute a description. Each description is combined with a small portion of support set and then combined data are sent to a cloud. With respect to decoding, upon receiving the request from a user, each cloud implements SRDS by using the acquired description. After reconstruction, the signal is still in the encrypted form, which protects the privacy of the signal. After receiving the reconstructed signal together with the support set, the user completes the decryption operation.

4.3.1 Secure SRDS in Multi-clouds

4.3.1.1 Service and Threat Model

Figure 4.2 depicts the service model for the architecture of secure SRDS in public multi-clouds. The basic message flow is as follows. The client firstly encrypts the 2D sparse signal with a secret key known to the user and then performs robust coding for the encrypted signal using parallel CS. Afterwards, the generated measurements are

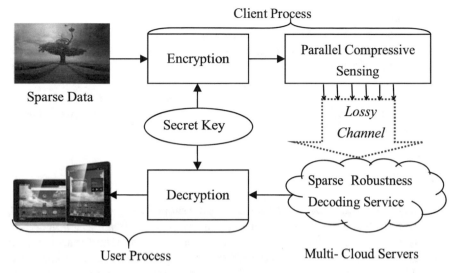

Fig. 4.2 The secure SRDS architecture

packeted and outsourced to the multi-cloud servers for storage and SRDS through packet-loss transmission. Once upon receiving the request from the user, the multi-clouds will fulfill the SRDS in parallel. Threats might come from multi-clouds as the adversary, so the multi-clouds are assumed to be semi-trusted, i.e., they honestly complete SRDS as specified but are curious about the client's data content. In addition, we also assume that the multi-clouds cannot collude with each other in private.

4.3.1.2   Design Goals

The proposed secure SRDS aims to achieve the following goals:

- Correctness
  The scheme can fulfill the task of SRDS outsourcing. The user can indeed receive a satisfactory sparse signal, provided that the client and the multi-clouds both follow the protocol honestly.
- Privacy
  The scheme should assure the input privacy and output privacy of the client's sparse signal, i.e., the multi-clouds cannot get meaningful information of the client's signal during signal storage and after SRDS.
- Effectiveness
  The robust encoding and decoding over encrypted sparse signal should work out effectively. The encryption algorithm should not have a negative impact on or even can enhance the performance of compression and reconstruction, in contrast to the case of no encryption.

- Robustness
  The encoding scheme should be robust enough against pack-loss. Even though high packet-loss rate occurs, the user can also obtain an acceptable quality level of the sparse signal.
- Efficiency
  Local computation and storage savings on the client and user sides should be substantially gained. The encryption algorithm should not result in substantial computation augment on the cloud side.

4.3.1.3   Basic Framework

Given a 2D sparse signal $\mathbf{X}$, the client first gets the support set $\mathbf{s}_v$ of $\mathbf{X}$ and encrypts $\mathbf{X}$ with a shifting operator $\mathbf{p}_1$ and an altering operator $\mathbf{p}_2$, both of which have the length of $|\mathbf{s}_v|$, where $|\cdot|$ represents the corresponding size. These two operators aim at hiding the indices and amplitudes of the nonzero entries of $\mathbf{X}$, respectively. They are unknown to the multi-clouds but need to be shared with the user. The client encrypts the non-zero entries in $\mathbf{X}$ with $\mathbf{p}_1$ and $\mathbf{p}_2$ to obtain the encrypted signal $\mathbf{X}'$ followed by conducting parallel CS column by column with the same sensing matrix Φ known to both the client and the multi-clouds. The obtained compressive measurements $\mathbf{Y}$ and support set $\mathbf{s}_v$ are grouped and sent to the multi-clouds for storage and SRDS through packet-loss networks. When the multi-clouds fulfill the SRDS, the reconstructed signal $\mathbf{X}''$ is still in encrypted form so that the signal privacy is preserved. Each cloud cannot immediately observe the original signal $\mathbf{X}$ with only a small portion of $\mathbf{s}_v$ and $\mathbf{Y}$. The basic framework of secure SRDS contains a collection of six polynomial-time algorithms marked as $\Omega = (SupGet, KeyGen, SigEnc, ParCS, SRDS, SigDec)$.

- $SupGet\,(\mathbf{X}) \rightarrow$ A support set vector generation algorithm that is run by the client. It takes the original signal $\mathbf{X}$ as input and returns a support vector $\mathbf{s}_v$.
- $KeyGen\,(1^p) \rightarrow$ A key generation algorithm that is run by the client. It takes a security parameter p as input and returns an encryption operator $\mathbf{p} = \mathbf{p}_1 \parallel \mathbf{p}_2$.
- $SigEnc\,(\mathbf{X}, \mathbf{p}) \rightarrow$ A signal encryption algorithm performed by the client who encrypts the signal $\mathbf{X}$ with $\mathbf{p}$ and returns the encrypted signal $\mathbf{X}'$.
- $ParCS\,(\mathbf{X}', \Phi) \rightarrow$ A parallel CS algorithm running on the client side. It takes the encrypted signal $\mathbf{X}'$ and the sensing matrix Φ as inputs and returns the measurements $\mathbf{Y}$.
- $SRDS\,(\mathbf{Y}, \Phi) \rightarrow$ An algorithm running on the cloud side to provide the SRDS for the user. It takes the measurements $\mathbf{Y}$ and the sensing matrix Φ as inputs and returns the reconstructed signal $\mathbf{X}''$.
- $SigDec\,(\mathbf{X}'', \mathbf{p}) \rightarrow$ A signal decryption algorithm that is handled by the user who decrypts the signal $\mathbf{X}''$ with the encryption operator $\mathbf{p}$ and returns the recovered original signal $\tilde{\mathbf{X}}$.

4.3.1.4   Design Details

In the following, we illustrate the design details of the basic framework of secure SRDS.

SupGet. The client changes $\mathbf{X}$ into 1D signal $\mathbf{x}$ of length $n = M \times N$ according to the column mode from left and right and then gets the support vector $\mathbf{s}_v$ of $\mathbf{x}$, as shown in Algorithm 14. Each entry of $\mathbf{s}_v$ lies in the integer interval $[1, n]$. Note that, if the signal from the client is not sparse but compressible, then it will be replaced by its s-sparse approximation [6].

Algorithm 14 Support Vector Generation

Input: X
Output: $\mathbf{s}_v$
1: $\mathbf{x} \leftarrow \mathbf{X}$
2: **for** $i = 1$ to n
3:    **if** $\mathbf{x}(i) \neq 0$
4:        Write i to $\mathbf{s}_v$
5:    **end if**
6: **end for**

KeyGen. The client employs a random number generator as the security parameter p, which is responsible for generating $\mathbf{p}$, as instantiated in Algorithm 15.

Algorithm 15 Key Generation

Input: p and $U = \{1, 2, \cdots, n\}$
Output: $\mathbf{p}=\mathbf{p}_1 \parallel \mathbf{p}_2$
1: Set $\mathbf{p}_1 = \mathbf{0}_{s \times 1}$ and $\mathbf{p}_2 = \mathbf{0}_{s \times 1}$
2: **for** $i = 1$ to s
3:    Select a random integer j from U
4:    Set $\mathbf{p}_1(i) = j$
5:    Update U by $U \leftarrow U/\{j\}$
6:    Select a random number η $(0 < \eta < 1)$
7:    Set $\mathbf{p}_2(i) = \eta$
8: **end for**

SigEnc. The client carries out encryption on $\mathbf{x}$ by using $\mathbf{p} = \mathbf{p}_1 \parallel \mathbf{p}_2$ and the encrypted signal is $\mathbf{x}'$, as shown in Algorithm 16. Here, employing $\mathbf{p}_1$ will randomly rearrange the non-zero entries of $\mathbf{x}$, which is equivalent to a random permutation of all entries, thus it can also relax the RIP of parallel CS with overwhelming probability. Apparently, the utilization of $\mathbf{p}_1$ has lower computational complexity and needs less memory space than using random permutation matrices. On the other hand, with the help of $\mathbf{p}_2$, the values of the nonzero entries of $\mathbf{x}$ will be altered. Note that the utilization of $\mathbf{p}_2$ does not change the number of non-zero entries of $\mathbf{x}$, i.e., maintaining the original sparsity after encryption, thus it does not affect the RIP.

Algorithm 16 Signal Encryption

Input: $\mathbf{p}$, $\mathbf{s}_v$ and $\mathbf{x}$
Output: $\mathbf{x}'$ and τ
1: Calculate $\tau = \max (abs (\mathbf{x}))$
2: Set $\mathbf{x}' = \mathbf{0}_{n \times 1}$
3: **for** $i = 1$ to s
4:    $\zeta = abs (\mathbf{x} (\mathbf{s}_v (i)))$
5:    $\mathbf{x}' (\mathbf{p}_1 (i)) = \text{sgn} [\mathbf{x} (\mathbf{s}_v (i))] \cdot [\zeta + (\tau - \zeta)/\tau \times \mathbf{p}_2 (i)]$
6: **end for**

ParCS. Prior to sampling, the encrypted data $\mathbf{x}'$ will be partitioned into multiple column data with equal length followed by repetitive sampling per column with the same sensing matrix $\mathbf{\Phi}$ known to the cloud servers. The sensing matrix could be a Gaussian matrix [4], Bernoulli matrix [5] or other matrix. At first glance, performing CS over encrypted signal looks irrational, because the encryption operation often destroys the data correlation from which the multimedia coders like JPEG profit. However, the proposed encryption algorithm not only makes parallel CS over encrypted domain possible but also can enhance the compression and reconstruction performance. It is the mainstay of our model, which roots in the idea of random permutation relaxing the RIP with high probability. Without loss of generality, let the column length be M, i.e, the partition result can be expressed as a matrix $\mathbf{X}'$ of $M \times N$. The parallel CS process returns a $K \times N$ matrix $\mathbf{Y}$, as shown in Algorithm 17. Then $\mathbf{Y}$ can be scheduled to multi-clouds through packet-loss networks.

Algorithm 17 Parallel Compressive Sensing

Input: $\mathbf{\Phi}$ and $\mathbf{X}'$
Output: $\mathbf{Y}$
1: Calculate $\mathbf{Y} = \mathbf{\Phi} \mathbf{X}'$

SRDS. The multi-cloud servers fulfill the SRDS by solving a convex optimization problem, as illustrated in Algorithm 18. Each cloud updates the sensing matrix in real time according to the received packet-lost measurements and then solves the corresponding optimization problem. Likewise, the parallel CS reconstruction architecture has lower complexity than the traditional CS counterpart.

Algorithm 18 Sparse Reconstruction Service

Input: $\mathbf{\Phi}$ and $\mathbf{Y}$
Output: $\mathbf{X}''$
1: Solve the following convex optimization problem:
    $\min \left\| \mathbf{X}'' [j] \right\|_1$ $s.t.$ $\mathbf{Y}[j] = \mathbf{\Phi} \mathbf{X}'' [j]$, $j \in [1, N]$

SigDec. Upon receiving the reconstructed signal $\mathbf{X}''$, the user transforms $\mathbf{X}''$ into the corresponding 1D form $\mathbf{x}''$ and conducts decryption using Algorithm 19, where

$\mathbf{s}_v$ can be gathered from the multi-cloud servers. The result is $\tilde{\mathbf{x}}$, which is further converted back into the 2D form $\tilde{\mathbf{X}}$.

Algorithm 19 Signal Decryption.

Input: $\mathbf{p}, \mathbf{s}_v, \mathbf{x}''$ and τ
Output: $\tilde{\mathbf{x}}$
1: $\tilde{\mathbf{x}} = \mathbf{0}_{n \times 1}$
2: **for** $i = s$ down to 1
3:    $\eta = \mathbf{x}'' (\mathbf{p}_1 (i))$
4:    $\tilde{\mathbf{x}} (\mathbf{s}_v (i)) = \text{sgn} (\eta) \cdot \left[abs (\eta) - \mathbf{p}_2 (i) \right] / (1 - \mathbf{p}_2 (i) / \tau)$
5: **end for**

4.3.2 Performance Evaluation

4.3.2.1 Experiment Settings

In the experiments, the 2D DCT coefficients of five 512×512 images, including Lena, Baboon, Boat, Peppers and Goldhill, are chosen as the compressive signals. Each of these signals is sparsified by using the best s-term approximation, i.e., retaining the front larger coefficients while setting the left smaller ones to zeros in line with zigzag order. The default is to keep the 10% front nonzero coefficients. This approximation has ever led to a PSNR decrease for reconstructed images and therefore it is set as a benchmark with which the experimental results are compared. Especially, the basis pursuit algorithm, CVX optimization toolbox [7], is utilized to realize Algorithm 5, where the measurement matrix consists of i.i.d. ensembles yielding Gaussian distribution. The communication latency between the client/user and the multi-clouds is ignored. When the descriptions are transmitted to the multi-clouds, the packet loss occurs at random. After a cloud receives a lossy description, it updates the corresponding measurement matrix by deleting some rows that are mapped to the row numbers of the lost measurements in a description. The packet loss rate, or PLR, does not exceed 30% in reality [8].

4.3.2.2 PLR Versus Sampling Rate (SR)

We consider a packet-loss channel[1] with the sampling rate $SR = \alpha \, (0 < \alpha \le 1)$ and the packet loss rate $PLR = \beta \, (0 \le \beta < 1)$. The SR is formulized as $SR = K/M$. We first investigate the connection of SR and PLR. In fact, the occurrence of packet loss can be thought as a decrease of SR. That is, given the equal significance that each

---

[1]Generally, a packet-loss channel with $SR = \alpha$ and $PLR = \beta$ is equivalent to a lossless channel with $SR = \alpha \, (1 - \beta)$.

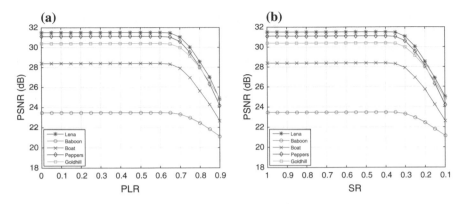

Fig. 4.3 **a** PSNR versus PLR with $SR = 1$; **b** PSNR versus SR with $PLR = 0$

packet carries, the PLR is proportional to the decrease of SR. This can be verified by comparing Fig. 4.3a and b, which show the PSNR versus PLR and SR, respectively. By visual comparison, we can see that the corresponding plots in Fig. 4.3a and b are almost identical and the same percentage of changes of PLR and SR have equal effect on the PSNR. This observation can help the client regulate the SR according to the PLR in real-time packet networks so that the reconstructed signal quality can be maintained at an acceptable level for the user.

Note that it is unnecessary to set very low SR as our focus is to conduct robust coding rather than compression coding for the client and the clouds have abundant resources for storage and computation. On the other hand, in the face of packet-loss networks, although the data to be transmitted should carry as much information as possible, a very high SR will waste transmission resources. As a result, the client should determine what SR is reasonable. This will be discussed in the subsection VI-D below.

4.3.2.3   Encryption Effectiveness

In this experiment, we inspect how the proposed encryption algorithm affect the performance of signal reconstruction, in comparison with the case of no encryption and a benchmark. The benchmark plot is generated by directly adopting the 10% front nonzero coefficients of the original signal for reconstruction. It can be seen from Fig. 4.4a–e that when PLR is smaller than 0.25 or so, both the cases of encryption and no encryption maintain the same reconstruction performance as the benchmark one. Then, with the increase of PLR, the case of no encryption has a continually declining PSNR. However, the case of encryption still sustains the same PSNR as the benchmark does until the PLR reaches about 0.65. With PLR varying from 0.65 to 0.9, the PSNR gain of the encryption case over the no-encryption case keeps up a rather stable but large margin. A detailed numerical comparison of the cases of

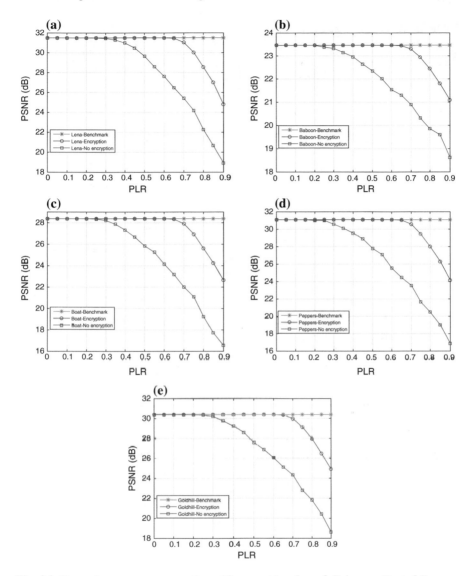

Fig. 4.4 Encryption versus no encryption with respect to **a** Lena, **b** Baboon, **c** Boat, **d** Peppers, and **e** Goldhill

encryption and no encryption on PSNR is shown in Table 4.1. It can be seen that the average PSNR gains for the five images are 5.81 dB, 2.41 (dB), 5.98 (dB), 7.26 (dB), and 5.95 (dB), respectively. In the meantime, we can also observe that with the increase of PLR, the PSNR gain will become increasingly great, meaning that the greater the PLR, the more effective the encryption will become.

Table 4.1 PSNR gain of the encryption case over the no-encryption case (dB)

Sparse data	PLR	No encryption	Encryption	PNSR gain	Average
Lena	0.65	26.48	31.46	4.98	5.81
	0.70	25.41	31.03	5.63	
	0.75	24.19	30.03	5.84	
	0.80	22.27	28.55	6.28	
	0.85	20.67	26.98	6.31	
	0.90	18.93	24.78	5.85	
Baboon	0.65	21.30	23.44	2.14	2.41
	0.70	20.89	23.30	2.41	
	0.75	20.32	22.95	2.63	
	0.80	19.86	22.45	2.59	
	0.85	19.60	21.80	2.20	
	0.90	18.62	21.09	2.47	
Boat	0.65	23.17	28.32	5.15	5.98
	0.70	22.02	27.91	5.89	
	0.75	21.10	26.95	5.85	
	0.80	19.24	25.61	6.37	
	0.85	17.73	24.26	6.53	
	0.90	16.55	22.63	6.09	
Peppers	0.65	24.46	31.04	6.58	7.26
	0.70	23.56	30.58	7.03	
	0.75	21.68	29.48	7.80	
	0.80	20.49	28.03	7.53	
	0.85	19.00	26.30	7.30	
	0.90	16.87	24.15	7.28	
Goldhill	0.65	25.14	30.34	5.20	5.95
	0.70	24.34	29.99	5.65	
	0.75	22.81	29.13	6.32	
	0.80	21.80	27.96	6.16	
	0.85	20.41	26.47	6.05	
	0.90	18.62	24.91	6.29	

The positive impact of encryption on signal reconstruction can also be verified by visually inspecting the reconstructed images in Fig. 4.5. The reconstruction performance enhancement is particularly deceptive when PLR is relatively large. The reason that the proposed encryption algorithm can enhance signal reconstruction performance stems from the theoretical finding that random permutation can efficiently relax the RIP of PCS with overwhelming probability. Luckily, random permutation is a key part of the proposed encryption algorithm.

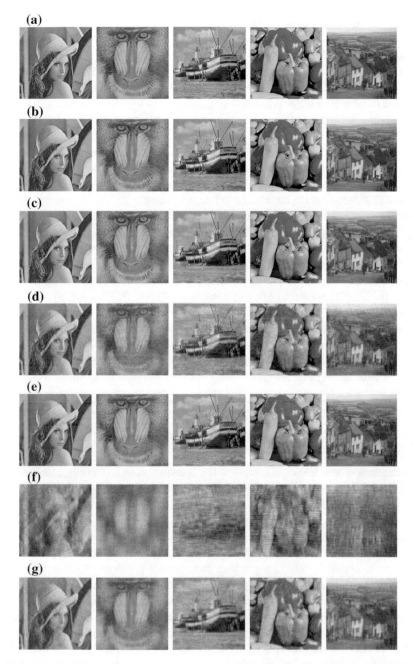

Fig. 4.5 The reconstructed images for different cases. **a** Benchmark, **b** No encryption, $PLR =$ 0.25, **c** Encryption, $PLR = 0.25$, **d** No encryption, $PLR = 0.65$, **e** Encryption, $PLR = 0.65$, **f** No encryption, $PLR = 0.90$, **g** Encryption, $PLR = 0.90$

4.3.2.4   Robustness Against Packet Loss

From Fig. 4.5a, c and e, we can find that the proposed encryption algorithm possesses
the identical visual effect as the benchmark one when PLR is no more than about
0.65. Even when PLR reaches 0.9, the outline is still visible on the whole. So the
proposed outsourcing scheme has a strong robustness against packet-loss networks.
The encoding way by the client can be viewed as efficient multiple description coding
using many descriptions to tackle packet loss.

   Now, let us have a look at robustness from a different perspective. As shown in
Fig. 4.4, the PSNRs of the no-encryption and encryption cases start to descend once
PLR exceeds approximately the transition points 0.25 and 0.65, respectively. For
the sake of generality, we denote these two transition points as γ_1 and γ_2, respectively. To attain satisfactory signal reconstruction quality for the user, by associ-
ating the relation of PLR and SR mentioned in the subsection VI-B above, we
have $\alpha(1 - \beta) \geq (1 - \gamma_i)$, where $i = 1, 2$. This means that given the no-encryption
case $(i = 1)$ or the encryption case $(i = 2)$, the client should select an SR that is
no less than $(1 - \gamma_i)/(1 - \beta)$. In practice, the PLR often does not surpass 30%
[8] and thus the SRs used in our experiment are $(1 - 0.25)/(1 - 0.3) \approx 1.07$ and
$(1 - 0.65)/(1 - 0.3) = 0.5$, corresponding to the cases without encryption and with
encryption, respectively. The SR value of 1.07 implies that even though there is no
compression in the no-encryption case, a good reconstruction performance same as
that of the benchmark case is not achievable. In short, the encryption algorithm in the
proposed outsourcing scheme can acquire stronger robustness, together with greater
compression rate, better reconstruction performance and privacy assurance.

4.3.2.5   Efficiency Evaluation

The efficiency of the proposed secure SRDS is tested using Lena image under different PLRs and the results are shown in Table 4.2, where the number of multi-clouds is
$l = 512$. Also, in Table 4.2, $t_{original}$ represents the time required to solve the original
reconstruction problem without encryption; $t_{customer}$, t_{cloud} and t_{user} are the time spent
by the client, the multi-clouds and the user, respectively, in the secure SRDS. After
encryption, the time of solving the reconstruction problem over encrypted domain
by the multi-clouds is often longer than the original problem and this *speeddown*
is reflected as $t_{cloud}/t_{original}$. The proposed secure SRDS offers the client or user
computation savings and this *speedup* is evaluated as $t_{original}/(t_{client} + t_{user})$.

   From Table 4.2, one can find by simple computation that the average speeddown
value is about 1.44, which means that the use of encryption causes a computation
increase of about 44%. Nevertheless, the advantages of using encryption have been
discussed in detail above. Importantly, from Table 4.2, it can be found that the average
speedup value is about 26, which means that the use of encryption has dramatically
reduced the computation cost of the client and/or user. We can also see from the
table that with the increase of PLR, the speedup value becomes smaller. At last,
we would like to note that we have also used various other natural images to test

Table 4.2 Efficiency test of secure SRDS using Lena image (time in seconds), where $l = 512$

No.	PLR	Original recovery	Secure SRDS			Speeddown	Speedup
		$t_{original}$	t_{client}	t_{cloud}	t_{user}	$\frac{t_{cloud}}{t_{original}}$	$\frac{t_{original}}{t_{client}+t_{user}}$
1	0.00	1204.9231	0.0445	1857.7857	0.0318	1.5418	30.8516
2	0.05	1075.1374	0.0142	1666.1122	0.0256	1.5497	52.8135
3	0.10	963.5391	0.0105	1463.3876	0.0246	1.5188	53.6953
4	0.15	888.3311	0.0096	1270.2471	0.0250	1.4299	50.1387
5	0.20	799.9329	0.0095	1136.3754	0.0254	1.4206	44.7578
6	0.25	766.4167	0.0097	1092.9472	0.0276	1.4260	40.1309
7	0.30	656.8616	0.0110	949.6108	0.0256	1.4457	34.9785
8	0.35	590.2177	0.0092	818.0449	0.0266	1.3860	32.2558
9	0.40	520.2159	0.0092	716.4444	0.0254	1.3772	29.3672
10	0.45	452.5533	0.0098	618.4225	0.0250	1.3665	25.3516
11	0.50	407.8672	0.0108	554.8367	0.0263	1.3603	21.4473
12	0.55	334.4333	0.0092	442.0381	0.0253	1.3218	18.9492
13	0.60	278.5116	0.0113	382.4820	0.0269	1.3733	14.2502
14	0.65	233.7896	0.0107	355.1916	0.0316	1.5193	10.7891
15	0.70	191.5983	0.0100	312.4333	0.0290	1.6307	9.5996
16	0.75	159.6331	0.0100	252.0124	0.0394	1.5787	6.3105
17	0.80	123.5063	0.0117	183.0870	0.0270	1.4824	6.2324
18	0.85	96.4469	0.0100	134.7367	0.0268	1.3970	5.1269
19	0.90	75.9973	0.0104	99.7831	0.0281	1.3130	3.8574

the proposed secure SRDS, in terms of PLR versus SR, encryption effectiveness, robustness against packet loss and efficiency evaluation. The experimental results are similar.

4.4   Concluding Remarks

This chapter discussed two CS reconstruction outsourcing protocols in multi-clouds. The first one is a privacy-preserving parallel outsourcing protocol, in which sparse reconstruction service is outsourced to multi-clouds that are curious about the signal content. The simple exchange primitive assures the security of support-set and is immune to the curiosity of multi-clouds. The parallel outsourcing mode has brought some benefits to the signal owner, the clouds and the user in terms of complexity, storage space and efficiency. The second one is a privacy-assured outsourcing protocol for SRDS through packet loss transmission. The encryption algorithm simply processes the indices and amplitudes of the non-zero entries in the 2D sparse signal.

Then the parallel CS is used for robust coding and the SRDS is outsourced to the multi-clouds, where each cloud only owns a small quantity of information of both the measurements and asymmetric support set. Upon the request of a user, the multi-clouds carry out SRDS and send the reconstructed signal, which is still encrypted, to the user. Finally, the user decrypts the reconstructed signal to recover the original signal. The superior performance of the proposed PAO-SRDS has been demonstrated by theoretical analysis and extensive experimental results.

References

1. M. AlZain, E. Pardede, B. Soh, J. Thom et al., Cloud computing security: from single to multi-clouds, in *Proceedings of 45th Hawaii International Conference on System Science, HICSS* (IEEE, 2012), pp. 5490–5499
2. H. Fang, S.A. Vorobyov, H. Jiang, O. Taheri, Permutation meets parallel compressed sensing: How to relax restricted isometry property for 2D sparse signals. IEEE Trans. Signal Process. **62**(1), 196–210 (2014)
3. C. Wang, B. Zhang, K. Ren, J. Wang, Privacy-assured outsourcing of image reconstruction service in cloud. IEEE Trans. Emerg. Top. Comput. **1**(1), 166–177 (2013)
4. E.J. Candès, T. Tao, Near-optimal signal recovery from random projections: Universal encoding strategies? IEEE Trans. Inf. Theory **52**(12), 5406–5425 (2006)
5. S. Mendelson, A. Pajor, N. Tomczak-Jaegermann, Uniform uncertainty principle for Bernoulli and subGaussian ensembles. Constr. Approx. **28**(3), 277–289 (2008)
6. E.J. Candès, M.B. Wakin, An introduction to compressive sampling. IEEE Signal Process. Mag. **25**(2), 21–30 (2008)
7. M. Grant, S. Boyd, Y. Ye, CVX: Matlab software for disciplined convex programming (2008)
8. J. Zhao, R. Govindan, Understanding packet delivery performance in dense wireless sensor networks, in *Proceedings of 1st International Conference on Embedded Network Sensor Systems* (2003), pp. 1–13

Chapter 5
Internet of Things Security

Abstract The CS's advantage of low-energy sampling makes it be well used for IoT with limited resources. Besides of energy consideration, this chapter focuses mainly on the security aspects. A low-cost and confidentiality-preserving data acquisition framework in IoT based on chaotic convolution and random subsampling is firstly proposed. Chaotic encryption ensures the security of sampling process. The sampled images are assembled into a big master image, which is encrypted by Arnold transform and single value diffusion. Both of these two encryption operations both have low computational complexity. The final encrypted image is uploaded to cloud servers for storage and decryption service. Then, we discussed the issue of how to securely store and share these big image data from IoT. We harness the hybrid cloud to provide secure big image data storage and share service for users. The basic idea is to partition each image into a small set of sensitive data and a large set of insensitive data, which are securely stored in the private cloud and the public cloud, respectively.

5.1 Compressive Sensing Meets Internet of Things

Internet of Things is a hot prospect now and is closely related to our life. One of challenges in IoT is to realize low-cost sampling and compression encoding, due to sensors' limited computation resources and large data volume. It is noteworthy that compression encoding can reduce the transmission bandwidth consumption and then save the power for the sensors. To address this challenge, CS has become a promising approach for data collection in IoT [1–3]. The advantage of CS is to simultaneously complete data sampling and compression based on the sparsity of the data to be sampled. With CS, a great deal of computation complexity has been shifted from the sampling side to the reconstruction side in the sense that the sampling complexity is linear in the dimension of the data while the reconstruction complexity is cubic. Such a favorable property of CS fits exactly the requirement of IoT, where resource-limited sensors communicate with a powerful data centre. In line with this observation, many works advocate the adoption of CS technology for IoT applications. Fragkiadakis et al. suggested an adaptive CS framework involving IoT applications in which a central smart object leverages a learning phase and provides feedback to the rest of the smart objects [1]. Li et al. investigated how CS can be used to perform data sampling and reconstruction for different information systems in IoT [2]. Dixon et al. considered some CS systems for electrocardiogram (ECG) and electromyogram (EMG) wireless biosensors by taking advantage of the sparsity of ECG and EMG signals [3].

5.2 Secure Low-Cost Compressive Sensing in Internet of Things

Internet of Things faces the challenge of how to realize low-cost data acquisition while still preserve data confidentiality. In the following, we present a low-cost and confidentiality-preserving data acquisition framework for IoT. Firstly, we harness chaotic convolution and random subsampling to capture multiple image signals. The measurement matrix is under the control of chaos, ensuring the security of the sampling process. Next, we assemble these sampled images into a big master image, and then encrypt this master image based on Arnold transform and single value diffusion. The computation of these two transforms only requires some low-complexity operations. Finally, the encrypted image is delivered to cloud servers for storage and decryption service. Experimental results demonstrate the security and effectiveness of the proposed framework.

5.2.1 Fundamental Techniques

5.2.1.1 Chaotic Convolution and Chaotic Subsampling

Random convolution can be considered as a kind of CS sampling theory that convolves the original signal with a random pulse [4]. The purpose of using the random pulse is to propagate the energy of the signal into the discrete spectrum uniformly. This process is formulate as

$$\mathbf{y} = 1 \Big/ \sqrt{N} \cdot \mathbf{F}^* \Omega \mathbf{F} \mathbf{x}, \tag{5.1}$$

where $\mathbf{F}$ is the discrete Fourier matrix, $\mathbf{F}^*$ is the inverse discrete Fourier matrix, and Ω is a diagonal matrix with the following form

$$\Omega = \begin{pmatrix} \sigma_1 \cdots 0 \\ \vdots \ddots \vdots \\ 0 \cdots \sigma_N \end{pmatrix}. \tag{5.2}$$

The diagonal elements of $\boldsymbol{\Omega}$, σ_ω, is subject to the following constraints: for $\omega = 1$, $\sigma_\omega = \pm 1$ with probability 1/2; for $2 \leq \omega < \frac{N}{2} + 1$, $\sigma_\omega = e^{j\theta_\omega}$, where $\theta_\omega \in [0, 2\pi]$ obeys the uniform distribution; for $\omega = \frac{N}{2} + 1$, $\sigma_\omega = \pm 1$ with probability 1/2; for $\frac{N}{2} + 2 \leq \omega \leq N$, $\sigma_\omega = \sigma^*_{N-l+2}$, where σ^*_{N-l+2} is the conjugate of σ_{N-l+2}.

Chaotic convolution means the random convolution is under the control of a chaotic system. Specifically, we choose the following cascade chaotic maps [5, 6] to instantiate our design

$$c_{i+1} = \begin{cases} v_1 v_2 c_i \left(1 - v_2 c_i\right), & c_i < 0.5 \\ v_1 v_2 \left(1 - c_i\right) \left(1 - v_2 \left(1 - c_i\right)\right), & c_i \geq 0.5 \end{cases}, \tag{5.3}$$

where $v_1 \in [3.57, 4]$ and $v_2 \in (1, 2]$. For a given initial key c_0^r, iterate Eq. (5.3) for $\nu \geq 1000$ times to avoid transient effect and further run Eq. (5.3) for $\lfloor N/2 - 1 \rfloor$ more times to get $c_1^r, c_2^r, \ldots, c_{\lfloor N/2-1\rfloor}^r$. We harness these values to build up $\boldsymbol{\Omega}$ as follows: if $\omega = 1$, set $\sigma_\omega = 1$; if $2 \leq \omega < \frac{N}{2} + 1$, set $\sigma_\omega = e^{j\theta_\omega}$, where $\theta_\omega = 2\pi c_\omega^r$; if $\omega = \frac{N}{2} + 1$, set $\sigma_\omega = -1$; if $\frac{N}{2} + 2 \leq \omega \leq N$, set $\sigma_\omega = \sigma^*_{N-l+2}$.

Chaotic subsampling is to randomly select some indexes in an image signal with the supervision of a chaotic system. Denote the sampling rate as τ, then there will be $M = N\tau$ locations sampled. Handle (5) with another initial key c_0^s, and produce M random integers $c_1^s, c_2^s, \ldots, c_M^s$ with the constraints that $c_i^s \in [1, N]$ and $c_i^s \neq c_j^s$ if $i \neq j$. These M random numbers can be utilized to construct an $M \times N$ chaotic subsampling operator ∇ by setting the i-th row ($i \in [1, M]$) of ∇ as $\nabla_i = [0, 0, \ldots, 0, 1_{c_i^s}, 0, \ldots, 0]$.

With the above notations, the whole CS process is described as

$$\mathbf{y} = 1 \big/ \sqrt{N} \cdot \nabla \left\{ \mathbf{F}^* \boldsymbol{\Omega} \mathbf{F} \mathbf{x} \right\}. \tag{5.4}$$

The chaotic convolution and chaotic-supervised random subsampling constitute a secure CS scheme, which is regarded as a symmetric cryptosystem, as shown in Fig. 5.1. The advantage of chaotic control is to save the space of storing the convolution matrix and subsampling matrices, which brings convenience for practical use in sensor devices.

5.2.1.2 Arnold Transform

Arnold transform can scramble the positions of the pixels in an image [7]. Thus, the correlation among the pixels of an image can be eliminated by transform the image several times. Arnold transform has the periodic property that if it is used to permute an image, then this image will be restored after a limited number of transformations. The period is related to the image size [7]. Assuming that the size of the original image is $N \times N$, Arnold transform can be described as follows:

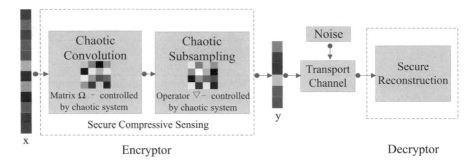

Fig. 5.1 A secure CS data acquisition scheme based on chaotic convolution and chaotic subsampling

$$\begin{bmatrix} X_{k+1} \\ Y_{k+1} \end{bmatrix} = \mathrm{mod}\left(\begin{bmatrix} 1 & b \\ a & ab+1 \end{bmatrix} \begin{bmatrix} X_k \\ Y_k \end{bmatrix}, N \right), \tag{5.5}$$

where $k \in \{1, 2, \ldots, N\}$, $(X_k, Y_k)^T$ denotes the positions of the original image pixels, and $(X_{k+1}, Y_{k+1})^T$ denotes the positions of image pixels after transformation. For example, if we set $a = b = 1$, then the above Arnold transform can be described as:

$$\begin{bmatrix} X_{k+1} \\ Y_{k+1} \end{bmatrix} = \mathrm{mod}\left(\begin{bmatrix} 1 & 1 \\ 1 & 2 \end{bmatrix} \begin{bmatrix} X_k \\ Y_k \end{bmatrix}, N \right), \tag{5.6}$$

where $k \in \{1, 2, \ldots, N\}$. The permutation can be reversed using the inverse Arnold transform, which is as follows:

$$\begin{bmatrix} X_k \\ Y_k \end{bmatrix} = \mathrm{mod}\left(\begin{bmatrix} 2 & -1 \\ -1 & 1 \end{bmatrix} \begin{bmatrix} X_{k+1} \\ Y_{k+1} \end{bmatrix}, N \right), \tag{5.7}$$

where $k \in \{1, 2, \ldots, N\}$.

5.2.1.3 Single Value Diffusion

The single value diffusion indicates each pixel value is separately XORed by a secret key, mathematically displayed as

$$\zeta' = \zeta \oplus c, \tag{5.8}$$

where ζ and ζ' represent the plaintext and ciphertext, respectively. Here, c is the secret key, which is provided by a iterating cascading chaotic system with initial key c_0^d. The inverse process is easily obtained as

$$\zeta = \zeta' \oplus c, \tag{5.9}$$

which is used for decryption.

5.2.2 Low-Cost and Confidentiality-Preserving Data Acquisition

5.2.2.1 Basic Framework Design

The proposed framework supports batch image processing, i.e., multiple images are simultaneously sampled, compressed, fused and encrypted. From a high level point of view, the service flow of the design is as follows. At first, multiple image signals are separately or one-by-one convoluted by chaotic convolution in a sensor node and then each convoluted image is subsampled under the supervision of chaotic systems. The chaotic convolution and chaotic subsampling are comprised of secure CS, which achieves simultaneous sampling, compression and encryption, and can solve the first challenge problem. However, as we mentioned earlier, the CS-based approach alone is not secure enough due to its intrinsic linearity. Thus, to enhance the security and process multiple images, we exploit a fusion encryption technique in which these subsampled images are combined into a complete master image in some way and then this master image is further encrypted with common image encryption techniques. For the master image combination, we make use of the simple cascading. It is worth noting that other Zig-Zag scanning methods, even secret combination methods, can also be used according to the practical requirement. For image encryption techniques, we choose Arnold transform and single value diffusion, which represent the permutation and diffusion operations, respectively. As a result, this fusion encryption owns the confusion and diffusion properties, where an ideal image cipher should possess. The final encrypted image is uploaded to cloud servers for storage, decryption, and reconstruction, since the cloud servers have vast storage and computing resources. Fig. 5.2 gives a visual illustration of the basic framework design. In the following, we will elaborate the proposed encryption and decryption designs. The main steps of encryption and decryption designs are shown in Fig. 5.3.

5.2.2.2 Encryption Design

The encryption design includes the following steps.

a: Perform chaotic convolution and chaotic subsampling for the input n image signals $\mathbf{x}_i, i = 1, 2, \ldots, n$. The measurement process is expressed as

$$\mathbf{y}_i = \mathbf{\Phi}\mathbf{x}_i, \tag{5.10}$$

where $\mathbf{y}_i$ of size $(N/n) \times 1$ represents the measurement result of the i-th image signal $\mathbf{x}_i$.

Remark 1 In the proposed encryption design, the sampling rate is defined as $\tau = 1/n$, which is to keep a unified image size during the encryption process with the purpose of facilitating practical setting and usage. After sampling, each sampled image contains N/n entries and the n images have a total entry number of N. With

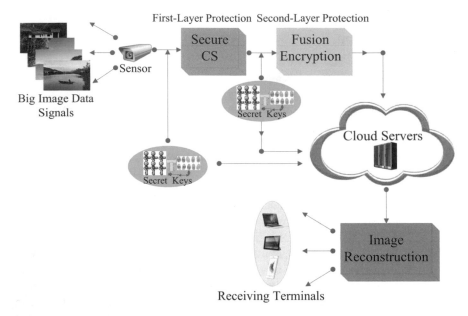

Fig. 5.2 A visual illustration of the basic framework design

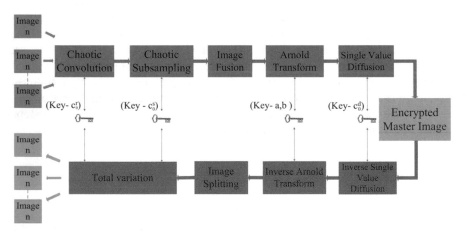

Fig. 5.3 Encryption and decryption designs

the increase of n, the compression performance is gradually enhanced, but each image after reconstruction declines in quality. As a consequence, it must have a balance between compression performance and image quality.

b. Cascade the sampled images y_i into a master image Y.

Remark 2 The cascading way can be selected according to practical requirement. For example, the simple one is to pile up images one after another. A slightly complex way is Zig-Zag scanning along some diagonal direction. In addition, some fixed templates or secret read way can be appropriately defined.

    c. Exploit Arnold transform to $\mathbf{Y}$ and mark the transformed result as $\mathbf{Y}_1$.

Remark 3 It is well known that Arnold transform used for permutation is not secure enough due to its inherent period. There are many permutation methods which have better permutation effect than Arnold transform. However, Arnold transform has very low computational complexity and is more suitable for resource-constrained sensors than other permutation methods. Thus, Arnold transform ensures a good trade-off between resource consumption and security guarantee.

    d. Diffuse $\mathbf{Y}_1$ using single value diffusion and mark the diffused result as $\mathbf{Y}_2$.

Remark 4 The XOR operation requires 8-bit integer type in the computing, therefore each entry in $\mathbf{Y}_1$ is quantized into the range [0, 255] using the sigmoid function [8] before single value diffusion. Similarly, the inverse sigmoid function is applied in the corresponding decryption process.

Remark 5 We carry out the single value diffusion, which is the simplest diffusion approach. In order for better diffusion effect, chain mode like Cipher Block Chaining or Cipher Feedback Mode often gets involved in diffusion [9–12]. Nevertheless, chain mode implies consuming more sensors' resources. A reasonable compromise between resource consumption and diffusion effect is the single value diffusion.

Remark 6 From the viewpoint of the whole implementation process, it can be easily found that the encryption design supports batch image signal processing. This caters to the processing of multimedia big data. Generally speaking, multiple image signals from the same sensor node will be regarded as a batch. This is because that the images captured by a sensor node often have a high similarity or correlation as well as strong redundancy if considered from a synthetic angle. A high degree of similarity or correlation can help users to quickly find out the related images of interest from the cloud servers. Strong redundancy means the further fusion and compression when they are stored on the cloud servers.

Remark 7 The encryption design utilizes two layers of encryption protection. The first layer is secure CS including chaotic convolution and chaotic subsampling. Chaos has high initial value sensitivity, which benefits secure sampling. The second layer is the fusion encryption with the permutation-diffusion structure. Double protection is convenient for the access control of image data in the cloud servers. Usually, the cloud servers are semi-trusted or malicious. When users send a remand, the second-layer protection can be straightforwardly decrypted by the cloud servers while the first-layer protection's decryption can be delivered to the trusted clouds or outsourced to semi-trusted clouds [13–15]. Experimental security analysis will be stated in the next section.

5.2.2.3 Decryption Design

The steps in the decryption design are as follows.

a. Perform the inverse single value diffusion for Y_2. The inverse of single value XOR can be easily calculated and the decrypted result is Y_1.

b. Exploit Arnold transform to Y_1. The decrypted result is Y.

c. Split Y into n images y_i, $i = 1, 2, \ldots n$.

d. Reconstruct n images x_i based on the total variation optimization algorithm [16, 17].

5.2.3 Simulation and Discussion

We select two image signals ($n = 2$), "Lena" and "Cameraman", which are used for testing. The image size is 256×256, i.e., N $= 256 \times 256 = 65536$. In image security field, some common security analyses such as histogram analysis, image entropy analysis, correlation analysis, and key sensitivity analysis are essential. Next, we perform these analyses together with robustness analysis, compression capability analysis and running time analysis.

5.2.3.1 Key Space Analysis

An excellent cryptosystem should be able to resist brute force attack, therefore a big key space is essential. In the proposed encryption framework, the keys include c_0^r, c_0^s, c_0^d, a and b. For the first three keys, if the key precision is 10^{16}, then the corresponding key space is at least 10^{48}. Together with the key space consisting of a and b in Arnold transform, on the whole, the key space is greater than 10^{48}. In fact, the key space can be further enlarged by setting v_1 and v_2 as new keys. Such a setting will result in six new keys, in three phases including chaotic convolution, chaotic subsampling, and single value diffusion shown in Fig. 5.3. Thus, we come to a conclusion that the proposed encryption framework has a very large key space against the exhaustive attacks.

5.2.3.2 Histogram Analysis

After completing encryption, the histogram of the encrypted image should greatly differ from that of the original one and even follow a uniform distribution so that it can withstand statistical analysis attack [18]. We perform the histogram analysis of adjacent pixels, as shown in Fig. 5.4. Figure 5.4a and b are the histograms of

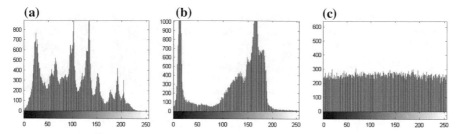

Fig. 5.4 **a** Histogram of Lena; **b** Histogram of Cameraman; **c** Histogram of encrypted image

the original Lena and Cameraman images respectively while Fig. 5.4c shows the histogram of the encrypted master image. Comparing Fig. 5.4a–c we can see that the histogram of the encrypted image has a huge difference from that of the original ones. Note that the histogram of the encrypted image can tend to the flat and uniform distribution if the more heavy image encryption techniques like multiple-round chain diffusion are used. However, the heavy encryption means more energy consumption. Hence, the present encryption techniques are ideal choices.

5.2.3.3 Image Entropy Analysis

Image entropy can quantitatively reflect the distribution characteristics of gray values, which can be defined as:

$$H = -\sum_{i=0}^{l-1} p(v_i)\log_2 p(v_i),\qquad(5.11)$$

where l is the gray level, v_i is the pixel value, and $p(v_i)$ is the probability of v_i. The sum of $p(v_i)$ is 1, i.e., $\sum_{i=0}^{l-1} p(v_i) = 1$.

If the pixel values approach to the uniform distribution, the entropy would get greater. The ideal case is that the probabilities of all encrypted pixel values are equal, then the encryption performance reaches the best. In this case, the image entropy is $\log_2 l$. When l=256, the ideal result is 8. Figure 5.5a and b are the entropies of the original images, i.e., 7.5730 and 7.1021, respectively. Figure 5.5c corresponds to the encrypted image, whose entropy is 7.9897, obviously closer to 8 in comparison with the original images' entropies. This shows the proposed scheme has a good encryption performance in terms of the distribution characteristics.

Fig. 5.5 Image entropies of the original images **a** and **b**, and the encrypted image **c**

(a) $H = 7.5730$ (b) $H = 7.1021$ (c) $H = 7.9897$

5.2.3.4 Correlation Analysis

The correlation coefficient is an important indicator to evaluate a cryptosystem. It can be defined as [19]

$$CC = \frac{\sum\limits_{i=1}^{N} (x_i - \bar{x})(y_i - \bar{y})}{\sqrt{\sum\limits_{i=1}^{N} (x_i - \bar{x})^2 \sum\limits_{i=1}^{N} (y_i - \bar{y})^2}}, \tag{5.12}$$

where $\bar{x} = 1/N \sum\limits_{i=1}^{N} x_i$ and $\bar{y} = 1/N \sum\limits_{i=1}^{N} y_i$. We randomly select 2000 pairs of pixels used for calculating the correlation coefficient in the horizontal, vertical and diagonal directions respectively, and the experiment results are shown in Table 5.1. It can be seen that for each original image, adjacent pixels generally have a strong correlation, but after encryption, the correlation of adjacent pixels has been reduced greatly. In addition, we depict the correlation distribution. Figure 5.6a–f corresponds to the original Lena and Cameraman images respectively, while Fig. 5.6g–i are the cases of the encrypted image, further confirming the above analysis.

5.2.3.5 Key Sensitivity Analysis

The chaotic system has the characteristic of high sensitivity to the initial parameters and values, which ensures that the proposed algorithm can provide a high level of key sensitivity. A visual verification is shown in Fig. 5.7, in which Fig. 5.7a and d

Table 5.1 Correlation coefficient of adjacent pixels

	Horizontal	Vercital	Diagonal
Lena (original)	0.9641	0.9273	0.9201
Cameraman (original)	0.9883	0.9823	0.9280
Encrypted image	0.0691	0.0564	0.0051

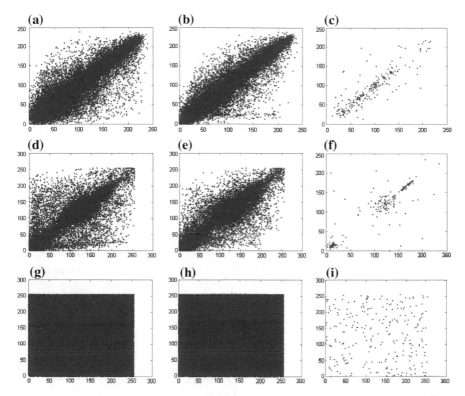

Fig. 5.6 Correlation distribution of adjacent pixels for **a** Lena image in horizontal direction; **b** Lena image in vertical direction; **c** Lena image in diagonal direction; **d** Cameraman image in horizontal direction; **e** Cameraman image in vertical direction; **f** Cameraman image in diagonal direction; **g** Encrypted image in horizontal direction; **h** Encrypted image in vertical direction; **i** Encrypted image in diagonal direction

are the decrypted Lena and Cameraman images with the correct keys, respectively. Figure 5.7b, c, e and f shows the cases of slightly disturbed keys with a disturbance level of 10^{-16}. One can see that a slight change leads to a completely incognizable decrypted image. For the purpose of a digitized verification, we introduce the mean square error (MSE) to evaluate the key sensitivity. Table 5.2 shows the MSE results for correct key and the error key, respectively. As shown in Table 5.2, even if a wrong key has only a slight deviation from the correct one, any useful information cannot be obtained from the original image. Hence, the proposed framework is highly sensitive to the keys. Our testing on other keys deduces the same conclusion.

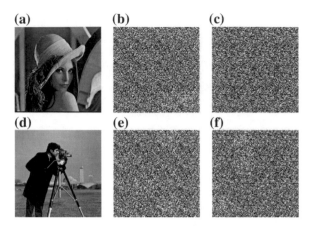

Fig. 5.7 Decrypted images with key sensitivity. **a** Lena image with correct key $c_0^r = 0.291$; **b** Lena image with a slightly disturbed Key $c_0^r = 0.291 + 10^{-16}$; **c** Lena image with a slightly disturbed key $c_0^r = 0.291 - 10^{-16}$; **d** Cameraman image with correct key $c_0^r = 0.291$; **e** Cameraman image with a slightly disturbed key $c_0^r = 0.291 + 10^{-16}$; **f** Cameraman image with a slightly disturbed key $c_0^r = 0.291 - 10^{-16}$

Table 5.2 MSE for correct and wrong keys

	Correct key-$c_0^r(0.291)$	Wrong key-$c_0^r(0.291 + 10^{-16})$	Wrong key-$c_0^r(0.291 - 10^{-16})$
Lena	7.7924	14473	14982
Cameraman	8.6931	20184	20260

5.2.3.6 Robustness Analysis

In the actual transport channel, the encrypted image is often mixed with noise. The encryption scheme must take full account of the effect of noise. An excellent encryption scheme is required to resist noise interference. In order to test the quality of the reconstructed image in the noisy environment, the typical Gaussian noise and Salt and Pepper noise are considered. For Gaussian noise, assume that the noise which is mixed in the encrypted image can be simulated as follows [20]:

$$y' = y + \alpha G, \tag{5.13}$$

where y is the encrypted image, α is the coefficient of noise interference, G represents the white Gaussian noise, and y' is the encrypted image mixed with noise interference. Figure 5.8a is the decrypted image at $\alpha = 0.1$, Fig. 5.8b is the decrypted image at $\alpha = 0.3$, and Fig. 5.8c is the decrypted image at $\alpha = 0.5$. In order to comprehensively analyze the influence of mixing with noise, Table 5.3 lists the PSNR and MSE results under different α. We can see that, the quality of the decrypted image declines with

Fig. 5.8 Decrypted images mixed with white Gaussian noise for **a** $\alpha = 0.1$; **b** $\alpha = 0.3$; **c** $\alpha = 0.5$

Table 5.3 PSNR And MSE For different noise coefficients (α)

	$\alpha = 0.1$	$\alpha = 0.2$	$\alpha = 0.3$	$\alpha = 0.4$	$\alpha = 0.5$
PSNR (dB)	39.5815	33.2447	21.1772	18.1745	15.4814
MSE	7.1269	30.8043	495.8641	989.9796	1840.5

Fig. 5.9 Decrypted images disturbed by Salt and Pepper noise. **a** 0.1% pixels are disturbed; b 0.5% pixels are disturbed; c 1% pixels are disturbed

the increase of α. In spite of this, the reconstructed image is still acceptable for a certain range of α.

For the Salt and Pepper noise, under its impact, the partial gray values of the image are changed into 0 or 255 randomly. Figure 5.9a–c shows the reconstructed image with 0.1%, 0.5%, and 1% pixels disturbed, respectively. In summary, the proposed encryption scheme can withstand a reasonable degree of typical noises.

5.2.3.7   Compression Capability Analysis

Reducing the sampling rate will improve the compression performance but inevitably lower the quality of the reconstructed image. The experiment results are showed in Fig. 5.10. From the results, the proposed scheme has a satisfactory compression performance. Even if the sampling rate is reduced to 1/8, the reconstructed image is still acceptable. It is rather remarkable that increasing the compression degree will lower the quality of the reconstructed image. It is necessary to set the appropriate sampling rate in practical applications. Furthermore, the camouflage image (without the actual value) can be added to the valuable images if necessary, which can play the role of disguise to further improve the security.

Original Images	Sample Rate	Decryted Images	PSNR(dB)	MSE

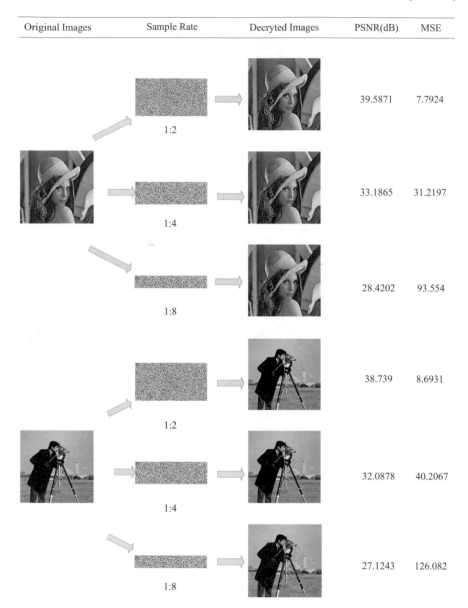

Fig. 5.10 PSNR and MSE under different sampling rates

5.3 Secure Data Storage and Sharing Service in Internet of Things

With the era of IoT, CCD/CMOS image sensors are pervasively deployed in everywhere such as digital camera, smartphone, traffic camera, satellite reconnaissance, astronomical observation, medical microendoscopy and robot vision. A great number of images are captured by these image sensors every day and the dramatically increasing number is drawing to people's concerns for big image data' storage and privacy. To protect the image privacy, people are accustomed to store the image data in the private cloud of their own, as the private cloud, in general, is credible. But with the expansion of big image data, the storage space in the private cloud is not vast enough. Thus, the private cloud has to seek help to the public cloud which possesses abundant storage and computing resources. However, the public cloud is often not trusted. As a result, the image data to be transmitted to the public cloud need to be encrypted. Meanwhile, compressing these encrypted image data appropriately before transmission is essential to reduce the consumption of communication resources. Nevertheless, it is not an easy task to exploit the image compression over encrypted domain [21]. On the other hand, it is highly expected that not all the data in an image are stored in the public cloud and a hand of data in an image, in which people are the most interested, can be placed in the private cloud for easy looking-up. Therefore, it is necessary to let the hybrid cloud provide an efficient image service while maintaining the privacy.

We design such an efficient secure image service framework for users through the hybrid cloud. Firstly, the private cloud divides an image into two parts, which are referred to as sensitive data and insensitive data, respectively. This division method can be selected by the private cloud which relies on the users' interests and requirements. In our work, we take the image's edge and contour for an example. An edge detector method is selected and used for distinguishing the sensitive data and the insensitive data. The sensitive data account for a tiny percentage (less than 20%) in an image while the remaining data are considered insensitive. Secondly, the insensitive data are appropriately encrypted and then compressed using compressed sampling technique. The compressed measurements are transmitted to a public cloud for storage. In addition, the sensitive data are directly encrypted and stored in the private cloud. Finally, once receiving a user's request, the public cloud exploits a privacy-preserving decompression and deliveries the decompressed data to the private cloud. The private cloud decrypts these decompressed data and encrypted sensitive data, and then assembles the decrypted results into a whole image for the user.

5.3.1 Secure Big Image Data Service Framework in the Hybrid Cloud

Different kinds of image sensors in IoT capture a lot of images every day. There exist an urgent problem of how to securely store these image data and efficiently serve for users. For security consideration, they need to be stored in the trusted private cloud and are encrypted appropriately, avoiding the illegal user to directly access the plain data. Meanwhile, the appropriate compression must be exploited to reduce storage space. However, the private cloud itself does not necessarily own such storage and computation resources with the era of big image data. On the other hand, the public cloud possesses more abundant storage and computation resources compared to the private cloud. As a result, it is desirable that the private cloud can leverage the public cloud's resources to serve for itself.

To make use of the public cloud's resources, the private cloud firstly partitions an image into two parts including sensitive data and insensitive data. The sensitive data that are of great significance only account for a very small percentage (e.g., no greater than 20%) of whole image data. The remaining are insensitive data (or called insignificant data). The sensitive data are straightway encrypted and stored in the private cloud side. For the insensitive data, the instinctive idea is to compress them and then transmit the compressed result to the public cloud for storage and decompression. Yet, the public cloud is often semi-trusted or even malicious and therefore it is required that the public cloud can fulfill a privacy-guaranteed decompression operation. In other words, prior to transmission, the insensitive data will be encrypted moderately. The image service framework is illustrated in Fig. 5.11. When a user requests an image, the public cloud decompresses the insensitive data and returns them to the private cloud which decrypts the sensitive and insensitive data and assembles both to a complete image for the user.

The proportion of the sensitive data is no more than 20% to well utilize the storage resources of the public cloud. However, this does not mean that a smaller proportion is better because of two reasons. One is that the sensitive data outperform the insensitive data in terms of security protection level, therefore, for the security purpose, as many sensitive data as possible should be stored in private cloud. The other is that processing the insensitive data is lossy. If too many data are partitioned into the insensitive part, then some important information in the image will be discarded. The sensitivity threshold value can be adjusted according to a practical setting. In general, 20% is a reasonable trade-off choice.

In the following, we give a specific realization based on the techniques of SED, Tent-Logistic system and compressed sampling. An image is firstly distinguished to obtain sensitive data set and insensitive data set with the help of SED. Then, the sensitive data are encrypted in parallel in the similar way of the counter mode. The insensitive data are encrypted with the architecture of permutation-diffusion and are then subsampled by parallel compressed sampling technique. The encrypted sensitive data are stored in the private cloud while the compressed measurements of the insensitive data are outsourced to the public cloud for storage. Such an outsourcing means the private cloud saves abundant storage space through outsourcing the greater than 80% data of an image to the public cloud. Upon receiving the user's request, the public cloud decompresses the compressed measurements and transmits the results that, in fact, are still encrypted insensitive data to the private cloud which decrypts and assembles the sensitive data and insensitive data into a complete image for the user. The encryption of insensitive data can be against ciphertext-only attack while

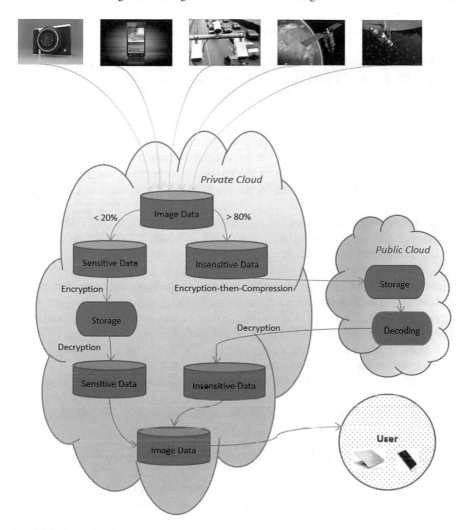

Fig. 5.11 Secure big image data service framework

the encryption of sensitive data can resist chosen-plaintext attack due the greater significance of the sensitive data than the insensitive data. It is worth mentioning that the proposed image service framework, as shown in Fig. 5.11, can be probably achieved using other more excellent techniques and methods, which will be further studied in future works. For example, the sensitive data can be further compressed before or after encryption, and the compressed measurements in the public cloud can also be considered to be compressed for storage space savings.

5.3.1.1 Data Sensitivity Identification in the Private Cloud

Sensitivity identification aims at extracting the sensitive data in an image. The sensitive data are the most interested region for users and limited to a very small proportion of an image. In this work, we choose the contour information of an image as region of interest. The steep region in contour change is more significant than the smooth region. To recognize these contour features, we can apply edge detection technique that is a common image processing method. An edge detection method of SED [22] is adopted to identify the location of sharp intensity transitions in a target image, finally acquiring a detection image for the target image. Implementing this detection image recognizes the sensitive data for the target image. The specific algorithm steps are as follows.

a. Apply Sobel edge detector to a target image $\mathbf{I}$ of size $\sqrt{n} \times \sqrt{n}$ and generate a binary detection image $\mathbf{I}'$ with the same size as $\mathbf{I}$, in which "1" and "0" represent the detected and undetected pixels, respectively.

b. The region with more number of "1"s means more great significance than that with less number of "1"s in the detection image $\mathbf{I}'$. To accurately and expediently find out the sensitive region, we partition the target image into some non-overlapping blocks, each containing $\sqrt{N} \times \sqrt{N}$ pixels. These blocks are successively marked as $I_i, i = 1, 2, \ldots, n/N$, and n/N is the total number of blocks.

c. For each block I_i, calculate the number α_i of "1"s from the corresponding block in the detection image and define the sensitivity level $\beta_i = \alpha_i/N$.

d. Set the threshold value $\gamma \in [0, 1]$ and then generate a binary sensitivity indicator $\mathbf{\Delta}$ as follows:

$$\Delta_i = \begin{cases} 1, & if \ \beta_i \geq \gamma \\ 0, & else \end{cases}. \tag{5.14}$$

e. If $\Delta_i = 1$, then the ith block is sensitive, else it is insensitive. The sensitive data set is extracted as $D^{sen} = \{I_1^{sen}, I_2^{sen}, \ldots, I_{N^{sen}}^{sen}\}$, where N^{sen} represents the number of sensitive blocks. The remaining are the insensitive data set, i.e., $D^{ins} = \{I_1^{ins}, I_2^{ins}, \ldots, I_{N^{ins}}^{ins}\}$, where N^{ins} is the number of insensitive blocks and $N^{sen} + N^{ins} = n/N$. Note that, the binary sensitivity indicator is stored in the private cloud for final assembling use. The size of the indicator is controllable. If an image has a quite large size, then through increasing the number of pixels in each block, the size of the indicator can be reduced. Generally speaking, only a bit for each block is acceptable to store in the private cloud, just like the standard image coding schemes that have some auxiliary for decoding use. The above sensitivity identification method is similar to our previous work [23].

5.3.1.2 Sensitive Data Encryption in the Private Cloud

For sensitive data (<20%) encryption, on the one hand, we hope the computation resources in the private cloud to be efficiently used. On the other hand, because of the significance of sensitive data, they need to be heavily encrypted to be against some potential attacks, such as known/chosen-plaintext attack and cipher-only attack, especially chosen-plaintext attack, which is a frequent and powerful attack. Thus, we apply the idea of counter mode, which is a common block cipher operation mode, as the counter mode has the advantage that encryption and decryption can be fully parallelized. In addition to the parallelism, it can resist chosen-plaintext attack under an appropriate keystream generator. Here, this generator is constructed by using the Tent-Logistic system [5], which is a new cascade chaotic map with the input of two one-dimensional seed maps, i.e., Logistic map and Tent map. Compared with two seed maps, the Tent-Logistic system possesses more chaotic performance, mathematically described as

$$z_{i+1} = \begin{cases} \delta_1 \delta_2 z_i (1 - \delta_2 z_i), & if \ z_i < 0.5 \\ \delta_1 \delta_2 (1 - z_i) (1 - \delta_2 (1 - z_i)), & else \end{cases}, \tag{5.15}$$

where the parameters $\delta_1 \in [3.57, 4]$ and $\delta_2 \in (1, 2]$.

Let $IK^{sen} \in (0, 1)$ be an initial key and assign secret values for δ_1^{sen} and δ_2^{sen}. The sensitive data set $D^{sen} = \{I_1^{sen}, I_2^{sen}, \ldots, I_{N_{sen}}^{sen}\}$ is encrypted using the following steps. Note that, there are two subscripts in the I_{j_1, j_2}^{sen}, which indicate the j_1th ($1 \leq j_1 \leq N_{sen}$) sensitive block and the j_2th ($1 \leq j_2 \leq N$) pixel in a block, respectively.

a. Initial value preparation. Calculate the initial value $z_{j_1, 0}$,

$$z_{j_1, 0} = \left(IK^{sen} + j_1/N^{sen}\right) \bmod 1, \tag{5.16}$$

where mod is the module operation.

b. Keysteam generation. The whole process is marked as a function $Z(\cdot)$. Firstly, iterate (2) with $z_{j_1, j_2 - 1}$, δ_1^{sen} and δ_2^{sen}, to output z_{j_1, j_2}. Then, map z_{j_1, j_2} into $Z_{j_1, j_2} \in [0, L]$, where L is the image gray level, using the formula

$$Z_{j_1, j_2} = floor\left(z_{j_1, j_2 - 1} \times 10^{14}\right) \bmod L, \tag{5.17}$$

where $floor(\cdot)$ is a function of the number rounded down. This step is expressed as, in short, $Z_{j_1, j_2} = Z\left(z_{j_1, j_2 - 1}\right)$.

c. Sensitive data encryption. Encrypt the pixel I_{j_1, j_2}^{sen} by

$$C_{j_1, j_2}^{sen} = I_{j_1, j_2}^{sen} \oplus Z_{j_1, j_2}, \tag{5.18}$$

where $\oplus$ is the exclusive or operation and C_{j_1, j_2}^{sen} means the corresponding ciphertext. Repeating Steps (2–3) until the whole pixels are encrypted. For better understanding, a graphic illustration is Fig. 5.12.

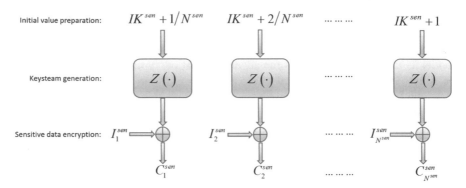

Fig. 5.12 An illustration of sensitive data encryption

5.3.1.3 Insensitive Data Encryption-then-Subsampling in the Private Cloud

The insensitive data with the amount exceeding 80% in an image will be compressed and then delivered to the public cloud for storage and decoding. However, the data need to be encrypted prior to transmission, since the public cloud is often semi-trusted or even malicious. Meanwhile, the computation resources in the public cloud side can be utilized to fulfill the decoding work. Thus, the private cloud performs the encryption then compression work before transmitting the insensitive data to the public cloud. So far, there have emerged some encryption then compression works [24–27], but there still existed an insurmountable issue, i.e., the incompatibility between encryption and compression. Encryption must affect the compression performance and thereby it is always making a trade-off by maximizing the compression performance only under two lightweight encryption schemes including permutation-only operation and diffusion-only operation. Such a permutation-only or diffusion-only operation is not secure enough and thus application-limited, since the permutation-diffusion operation is the fundamental architecture in image encryption field [6, 10, 11].

We consider the problem of whether or not it is possible to kill two birds with one stone rather than get half the result with twice the effort. The answer is yes. It is possible to enhance the security without affecting the compression performance by use of the permutation-diffusion architecture. Our work applies the idea that the appropriate encryption operation can be compatible with parallel compressed sampling.

Traditional compressed sampling [28, 29] samples a 1D sparse signal or compressible signal using a measurement matrix and can well recovery the signal from a few linear projections through utilizing a sensing matrix and solving a optimization problem. For a 2D image signal, if transformed into 1D form, then it will force the size of the sensing matrix in the reconstruction to become extremely large and the storage and computational complexity to increase dramatically. To address this problem, a common solution is to partition the image signal into multiple 1D sub-signals and simultaneously sample each sub-signal with the same measurement matrix. Obviously, this sampling can be achieved in parallel, referred to as parallel compressed sampling [30]. A natural partition is each column regarded as a sub-signal for an image.

Two crucial criterions in compressed sampling theory are the sparsity of the signal and the restricted isometry property of the sensing matrix. The introduced encryption operation should at least follow or even benefit these two criterions. It has been demonstrated that random permutation operation not only keeps the sparsity of the signal, but also benefits the restricted isometry property in parallel compressed sampling [31, 32]. The benefit can be understood that after random permutation, the sparsity of each column keeps a similar level and sampling each column with the same sampling rate makes the reconstruction performance better. This idea of non-zero element dispersion distribution can guide us to design other permutation operations. Meanwhile, the slight change of non-zero elements is used to realize the diffusion operation, as it has no impact on two criterions. After implementing permutation-diffusion encryption, we sub-samples each column using a measurement matrix to achieve the compression effect.

Although the compression performance of compressed sampling has been found not excellent enough [33], it possesses some outstanding advantages of its own that are concluded as the following several aspects. First of all, the insensitive data can be extremely sparse, which is very suitable for compressed sampling. The texture change in the insensitive data block is less apparent than that in the sensitive block. As a consequence, the corresponding non-zero elements after sparse transform have less number and the sparsity is stronger. Secondly, one can adopt a very small sampling rate to promote the compression performance, since the insensitivity means the insignificance, as claimed by the insensitive block itself. Thirdly, for security guarantee purpose, the permutation-diffusion architecture can be utilized, which is not only compatible with parallel compressed sampling, but also reinforces the compression performance. Moreover, the proposed compression scheme is robust and can offer progressive image reconstruction service. The robustness means on one hand, it can sustain the occurrence of packet-loss in the transmission channel between the private cloud and the public cloud. On the other hand, once partial data are returned by the public cloud, the private cloud is still able to provide progressive image service for users, as long as the reconstructed image quality is acceptable. Lastly, what we want to emphasize is not maximum compression but enough compression (subsampling). The compression data are sent to the public cloud for storage and decoding while maintaining the privacy. When required, the data are decoded by the public cloud and then returned back to the private cloud.

With respect to the insensitive data encryption-then-subsampling, the specific algorithmic procedures are stated as:

a. Insensitive matrix preparation. Perform DCT2 for each block in the insensitive data set $D^{ins} = \left\{ I_1^{ins}, I_2^{ins}, \ldots, I_{N^{ins}}^{ins} \right\}$ and rearrange the generated coefficients into a matrix $\mathbf{D}^{ins}$ of size $N \times N^{ins}$, where the ith column is from the ith block.

b. Index matrix generation. Set the initial value $IK_1^{ins} \in (0, 1)$ and the parameters δ_1^{ins} and δ_2^{ins} and iterate (2) to get a matrix $\mathbf{T}_1$ sized $N \times N^{ins}$, which is then mapped into an index matrix $\mathbf{Inx}$, in which each entry lies in the interval $(1, N')$, where $N' = N \cdot N^{ins}$. The mapping rule is to rearrange the elements in $\mathbf{T}_1$ in descending order and each entry in $\mathbf{Inx}$ represents one of locations of the elements in $\mathbf{T}_1$.

c. Permutation. Permute $\mathbf{D}^{ins}$ to obtain $\mathbf{D}_p^{ins}$ according to $\mathbf{Inx}$,

$$\mathbf{D}_p^{ins} (i, j) = \mathbf{D}^{ins} (i', j'), \tag{5.19}$$

where

$$i' = \begin{cases} N, & if \ \mathbf{Inx} (i, j) \bmod N = 0 \\ \mathbf{Inx} (i, j) \bmod N, & else \end{cases},$$

and

$$j' = ceil \left(\mathbf{Inx} (i, j)/N \right),$$

in which $ceil \, (\cdot)$ is a function of the number rounded up.

d. Perturbation matrix generation. Set another initial value $IK_2^{ins} \in (0, 1)$ and iterate (2) to acquire a matrix $\mathbf{T}_2$ sized $N \times N^{ins}$.

e. Diffusion. Diffuse $\mathbf{D}_p^{ins}$ to obtain $\mathbf{C}^{ins}$ according to $\mathbf{T}_2$,

$$\mathbf{C}^{ins} = \mathbf{D}_p^{ins}. \times ((\mathbf{T}_2 + 99.5)/100), \tag{5.20}$$

where $.\times$ represents the dot product.

f. Subsampling. Sample each column in $\mathbf{C}^{ins}$ to acquire the compressed measurements $\mathbf{Y}$ with a known measurement matrix $\mathbf{\Theta}$ of size $M \times N$ ($M < N$) known to both the private cloud and the public cloud.

$$\mathbf{Y} = \mathbf{\Theta} \mathbf{C}^{ins}. \tag{5.21}$$

Figure 5.13 gives an intuitive understanding for the insensitive data encryption.

5.3.1.4   Privacy-Guaranteed Insensitive Data Reconstruction in the Public Cloud

After generating the compressed measurements $\mathbf{Y}$, the private cloud outsources them to the public cloud for storage and decoding. Upon receiving the request from the private cloud, the public cloud exploits insensitive data reconstruction service in parallel as follows:

$$\min \left\| \tilde{\mathbf{C}}^{ins} \langle l \rangle \right\|_1 \ s.t. \ \mathbf{Y} \langle l \rangle = \mathbf{\Theta} \tilde{\mathbf{C}}^{ins} \langle l \rangle, \ l = 1, 2, \ldots, N_{ins}, \tag{5.22}$$

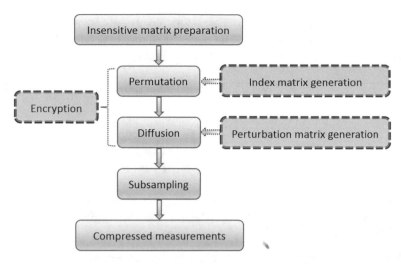

Fig. 5.13 An illustration of insensitive data encryption-then-subsampling

where $\tilde{\mathbf{C}}^{ins} \langle l \rangle$ means the lth column in $\tilde{\mathbf{C}}^{ins}$. The public cloud is assumed to be semi-trusted, i.e., can honestly fulfill the reconstruction service as specified, but is curious about the insensitive data. Obviously, if the privacy of $\tilde{\mathbf{C}}^{ins}$ can be guaranteed, so can the privacy of $\mathbf{Y}$, as $\mathbf{Y}$ is the compressed measurements of the encrypted data $\mathbf{C}^{ins}$. The privacy of $\tilde{\mathbf{C}}^{ins}$ is achievable, as proved by the next section, therefore the public cloud cannot acquire meaning knowledge from the insensitive data. The encryption complexity is $\mathcal{O}\left(\left(N^{ins}\right)^{2}\right)$ due to the ranking method during the index matrix generation phase while the reconstruction complexity is $\mathcal{O}\left(\left(N^{ins}\right)^{3}\right)$. It shows the outsourcing can gain the computation savings for the private cloud. If the greater computation savings is required in the private cloud, then a linear encryption complexity is achievable at an expense of removing the permutation encryption.

5.3.1.5   Image Assembling for Users in the Private Cloud

If receiving a request for some image from a user, the private cloud exploits three steps to serve for the user.

a. Decrypts the sensitive data by using the following equation:

$$I_{j_1,j_2}^{sen} = C_{j_1,j_2}^{sen} \oplus Z_{j_1,j_2}, \tag{5.23}$$

where Z_{j_1,j_2} is generated by the same Steps 1 and 2 in the encryption algorithm.

b. Decrypts the insensitive data. The public cloud provides the reconstruction service to send $\tilde{\mathbf{C}}^{ins}$ to the private cloud, which then performs inverse diffusion and inverse permutation operations as the following forms:

$$\tilde{\mathbf{D}}_p^{ins} = \mathbf{C}^{ins}./ \left((\mathbf{T}_2 + 99.5)/100 \right), \qquad (5.24)$$

and

$$\tilde{\mathbf{D}}^{ins} \left(i', j' \right) = \tilde{\mathbf{D}}_p^{ins} \left(i, j \right), \qquad (5.25)$$

where ./ represents the dot division. i', j', and $\mathbf{T}_2$ are the same as the encryption algorithm.

c. Regroup the image. Through performing the above two steps, the sensitive data set $D^{sen} = \left\{ I_1^{sen}, I_2^{sen}, \ldots, I_{Nsen}^{sen} \right\}$ and the insensitive data set $\tilde{D}^{ins} = \left\{ \tilde{I}_1^{ins}, \tilde{I}_2^{ins}, \ldots, \tilde{I}_{Nins}^{ins} \right\}$ are obtained, respectively. These two data sets are then regrouped into a complete image according to the corresponding binary sensitivity indicator Δ.

5.3.2 Performance Evaluation

5.3.2.1 Experiment Settings

We select Peppers image sized 512×512 as the test image used in our experiment. It will be partitioned into non-overlapping blocks with each one containing 64 pixels and then the total number of the blocks is 4096. The threshold value is selected as $\gamma = 0.05$. The keys are optionally set as $IK^{sen} = 0.35666852$, $\delta_1^{sen} = 3.99568435$, $\delta_2^{sen} = 1.99584626$, $IK_1^{ins} = 0.78995268$, $\delta_1^{ins} = 3.98546813$, $\delta_2^{ins} = 1.98567878$ and $IK_2^{ins} = 0.35946114$. To avoid chaotic temporary effect in the Tent-Logistic system, the first 1000 values are discarded. The sampling rate is chosen as $SR = 0.5$ and the measurement matrix Θ is comprised of independent identically distributed ensembles yielding Gaussian distribution. The convex optimization toolbox in [34] is used for insensitive data reconstruction in the public cloud.

5.3.2.2 Key Space Security Analysis

Regarding a good cryptosystem, the corresponding key space should be great enough to resist the brute-force attack. In the proposed cryptosystem, there are seven keys in total. If the key precision is 16, then the key space will be about 10^{112}, which can satisfy the requirement of resisting the brute-force attack. In fact, the key space can be further enlarged by adopting new values for the parameters δ_1 and δ_2 during the perturbation matrix generation phase when encrypting-then-subsampling the insensitive data. Note that, if an attacker wants to reveal either the sensitive data or the insensitive data, then he/she only needs to exhaust partial key space. For the sensitive data, the key space seems not sufficiently great due to only having three keys IK^{sen}, δ_1^{sen} and δ_2^{sen}. However, the encryption mode for the sensitive data is Counter type, which demonstrated secure against chosen-plaintext attack [35]. As for the insensitive data, the key space spanned by four keys IK_1^{ins}, δ_1^{ins}, δ_2^{ins} and IK_2^{ins} is in fact large enough against low-complexity exhaustive attacks and meanwhile, these four keys can be frequently altered. It is worth mentioning that in the proposed scheme, the binary sensitivity indicator Δ can be regarded as a key, whose possibility is 2^{4096}. Even if an attacker cryptanalyzes the whole blocks, he/she is still at a loss to regroup the original image without the knowledge of Δ.

5.3.2.3 Visual Security Analysis

The encrypted sensitive data are stored in the private cloud and the encrypted-then-subsampled insensitive data are placed in the public cloud. For the sake of intuitive understanding, the visual analyses of the encrypted results are depicted in Fig. 5.14, in which Fig. 5.14a, b and c are the visual effect of the original image, the encrypted sensitive data and the encrypted-then-subsampled insensitive data, respectively. Contrasting them can easily find that the encrypted data can cover the useful information of the original image. Also the encryption effect of the sensitive data is better than that of the insensitive data, since the former looks smoother than the latter from texture feature.

5.3.2.4 Key Sensitivity Security Analysis

The key sensitivity is pointed out that if a correct key slightly fluctuates, then it leads to that the decrypted data change dramatically. The stronger the sensitivity is, the better a cryptosystem is. It is well known that the parameter keys δ_1^{sen}, δ_2^{sen}, δ_1^{ins}, and δ_2^{ins} do not have the strong sensitivity, since they are fixed during the iteration process. Thus, we mainly investigates the sensitivity of two initial keys IK^{sen} and IK_1^{ins}. Specifically, we set a fluctuation of 10^{-16} with respect to three cases: (a) IK^{sen},

Fig. 5.14 The visual effect for **a** the original image; **b** the encrypted sensitive data; **c** the encrypted-then-subsampled insensitive data

Fig. 5.15 The sensitivity for
a IK^{sen}; **b** IK_1^{ins}; **c** IK^{sen}
and IK_1^{ins}

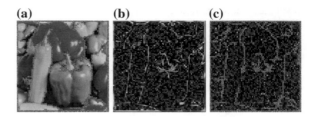

(b) IK_1^{ins}, and (c) IK^{sen} and IK_1^{ins}. The decrypted images are shown in Fig. 5.15, in which Fig. 5.15a reflects the sensitivity of IK^{sen} to the encrypted sensitive data, Fig. 5.15b indicates the sensitivity of IK_1^{ins} to the decrypted insensitive data, and Fig. 5.15c means the sensitivity of IK^{sen} and IK_1^{ins} to the whole decrypted data. As can be seen from Fig. 5.15, the sensitivity is very high.

5.3.2.5   Outsourcing Security and Encryption Reliability Analysis

The encryption of the insensitive data possesses the architecture of permutation-diffusion. In the diffusion operation, each entry is independently handled and no chain mode is adopted. Such one round permutation-diffusion is insecure against plaintext attacks. However, a heavy encryption will severely affect the subsampling and reconstruction effect. Thus, on one hand, the proposed encryption employs the session keys to resist plaintext attacks. On the other hand, the encryption reliability is verified in Fig. 5.16, since the case of the encryption promotes almost 2(dB) for the recovered image quality in comparison with the case of no encryption.

5.3.2.6   Compression Rate Versus Image Service Quality

In the proposed service framework, each image is compressed and distributed in the private cloud and the public cloud. The compression rate CR is related to two factors including the threshold value γ and the sampling rate SR. The threshold value γ determines what percentages of an image are stored in the private cloud and the public cloud, respectively. For example, in our experiment, $\gamma = 0.05$ results in $N^{sen} = 741$ and $N^{ins} = 3355$, and then two percentages are $N \cdot N^{sen}/n = 741/4096 \doteq 18\%$ and $N \cdot N^{ins}/n = 3355/4096 \doteq 82\%$. In general, the compression rate can be formulated as

$$CR = \frac{N(N_{sen} + N_{ins} \cdot SR)}{n}.$$

The compression rate has a direct impact on the regrouped image quality and we investigate their relationship in Fig. 5.16, where the image quality is measured by PSNR. Users can let the private cloud adjust the compression rate according to their image quality requirement. It is worth stressing that the compression performance is

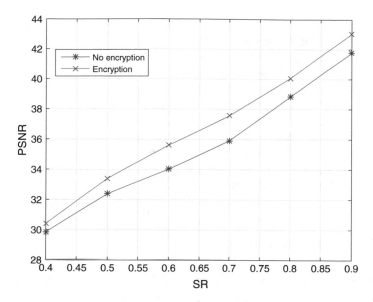

Fig. 5.16 Encryption reliability analysis

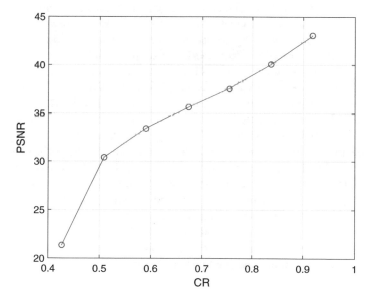

Fig. 5.17 Compression rate versus image service quality

not remarkable from Fig. 5.17, but our main goal is not the compression rate. What we want to do is let the private cloud only preserves a small percentage for each image and the most percentage is outsourced to the public cloud for storage. If the public cloud wants to save its storage resources, then it can compress the insensitive data by using other appropriate compression techniques. We do not further consider it in the present work. In addition, the value of γ depends mainly on the image's important degree and the private cloud's storage capability. If an image is thought of as more important, then reducing γ makes the private cloud store more sensitive data. Meanwhile, the private cloud's storage capability needs be also taken in account. Thus, there is a trade-off between the image's important degree and the private cloud's storage capability for the value of γ.

5.4 Concluding Remarks

This chapter considers IoT security based on CS. The first work is to design a low-cost and privacy-preserving data sampling approach in IoT. A double-layer protection mechanism based on chaotic encryption is embedded in the proposed approach. One is to embed chaotic convolution and chaotic subsampling during the sampling process and the other is to embed a permutation-diffusion structure after sampling. This two-layer mechanism can effectively protect data acquisition with very low computational cost. Meanwhile, the proposed framework can support batch big image data signal processing. Experimental results including key space analysis, histogram analysis, image entropy analysis, correlation analysis, and key sensitivity analysis demonstrate the security. Noise attack is verified through robustness analysis. The quality of the reconstructed images is validated by compression capability analysis. Running time analysis shows small encryption time. The proposed approach can have a potential application to IoT. The second work is the design of an efficient secure service framework for big image data with the help of the hybrid cloud. In an image, the sensitive data are securely stored in the private cloud while the insensitive data are encrypted-then-subsampled and stored in the public cloud. This framework can save at least 80% space for the private cloud. The public cloud is responsible for at least 80% the insensitive data's storage and decoding. The security is verified through key space analysis, visual analysis, and key sensitivity analysis. With respect to the insensitive data outsourcing, the encryption reliability is also demonstrated.

References

1. A. Fragkiadakis, P. Charalampidis, E. Tragos, Adaptive compressive sensing for energy efficient smart objects in IoT applications, in *4th International Conference on Wireless Communications, Vehicular Technology, Information Theory and Aerospace and Electronic Systems, VITAE* (IEEE, 2014), pp. 1–5

2. S. Li, L. Da Xu, X. Wang, Compressed sensing signal and data acquisition in wireless sensor networks and internet of things. IEEE Trans. Ind. Infor. **9**(4), 2177–2186 (2013)
3. A.M. Dixon, E.G. Allstot, D. Gangopadhyay, D.J. Allstot, Compressed sensing system considerations for ECG and EMG wireless biosensors. IEEE Trans. Biomedical Circ. Syst. **6**(2), 156–166 (2012)
4. J. Romberg, Compressive sensing by random convolution. SIAM J. Imag. Sci. **2**(4), 1098–1128 (2009)
5. Y. Zhou, Z. Hua, C.-M. Pun, C.P. Chen, Cascade chaotic system with applications. IEEE Trans. Cyber. **45**(9), 2001–2012 (2015)
6. Z. Hua, Y. Zhou, Image encryption using 2D Logistic-adjusted-Sine map. Inf. Sci. **339**, 237–253 (2016)
7. M. Khalili, D. Asatryan, Colour spaces effects on improved discrete wavelet transform-based digital image watermarking using Arnold transform map. IET Signal Process. **7**(3), 177–187 (2013)
8. Y. Li, B. Song, R. Cao, Y. Zhang, H. Qin, Image encryption based on compressive sensing and scrambled index for secure multimedia transmission. ACM Trans. Multimed. Comput. Commun. Appl. **12**(4s), 62 (2016)
9. L.Y. Zhang, Y. Liu, F. Pareschi, Y. Zhang, K.-W. Wong, R. Rovatti, G. Setti, On the security of a class of diffusion mechanisms for image encryption. IEEE Trans. Cybern. **48**(4), 1163–1175 (2018)
10. Y. Zhang, D. Xiao, Y. Shu, J. Li, A novel image encryption scheme based on a linear hyperbolic chaotic system of partial differential equations. Signal Process.-Image Commun. **28**(3), 292–300 (2013)
11. L.Y. Zhang, X. Hu, Y. Liu, K.-W. Wong, J. Gan, A chaotic image encryption scheme owning temp-value feedback. Commun. Nonlinear Sci. Numer. Simu. **19**(10), 3653–3659 (2014)
12. J.-X. Chen, Z.-L. Zhu, C. Fu, L.-B. Zhang, Y. Zhang, An efficient image encryption scheme using lookup table-based confusion and diffusion. Nonlinear Dyn., 1–16 (2015)
13. C. Wang, B. Zhang, K. Ren, J. Wang, Privacy-assured outsourcing of image reconstruction service in cloud. IEEE Trans. Emerg. Top. Comput. **1**(1), 166–177 (2013)
14. Y. Zhang, J. Zhou, Y. Xiang, L.Y. Zhang, F. Chen, S. Pang, X. Liao, Computation outsourcing meets lossy channel: secure sparse robustness decoding service in multi-clouds. IEEE Trans. Big Data (in press, 2017)
15. Y. Zhang, H. Huang, Y. Xiang, L.Y. Zhang, X. He, Harnessing the hybrid cloud for secure big image data service. IEEE Internet Things J. **4**(5), 1380–1388 (2017)
16. L.I. Rudin, S. Osher, E. Fatemi, Nonlinear total variation based noise removal algorithms. Phys. D Nonlinear Phenom. **60**(1–4), 259–268 (1992)
17. S.D. Babacan, R. Molina, A.K. Katsaggelos, Variational bayesian blind deconvolution using a total variation prior. IEEE Trans. Image Process. **18**(1), 12–26 (2009)
18. J.-X. Chen, Z.-L. Zhu, C. Fu, L.-B. Zhang, Y. Zhang, An image encryption scheme using nonlinear inter-pixel computing and swapping based permutation approach. Commun. Nonlinear Sci. Numerical Simu. **23**(1–3), 294–310 (2015)
19. Z. Hua, Y. Zhou, Design of image cipher using block-based scrambling and image filtering. Inf. Sci. **396**, 97–113 (2017)
20. N. Zhou, Y. Wang, L. Gong, H. He, J. Wu, Novel single-channel color image encryption algorithm based on chaos and fractional fourier transform. Opt. Commun. **284**(12), 2789–2796 (2011)
21. M. Johnson, P. Ishwar, V. Prabhakaran, D. Schonberg, K. Ramchandran, On compressing encrypted data. IEEE Trans. Signal Process. **52**(10), 2992–3006 (2004)
22. Wikipedia (2018), https://en.wikipedia.org/wiki/Sobel_operator
23. Y. Zhang, D. Xiao, W. Wen, Y. Tian, Edge-based lightweight image encryption using chaos-based reversible hidden transform and multiple-order discrete fractional cosine transform. Opt. Laser Tech. **54**, 1–6 (2013)
24. X. Zhang, Lossy compression and iterative reconstruction for encrypted image. IEEE Trans. Inf. Forensics Sec. **6**(1), 53–58 (2011)

25. X. Zhang, G. Feng, Y. Ren, Z. Qian, Scalable coding of encrypted images. IEEE Trans. Image Process. **21**(6), 3108–3114 (2012)
26. J. Zhou, X. Liu, O.C. Au, Y.Y. Tang, Designing an efficient image encryption-then-compression system via prediction error clustering and random permutation. IEEE trans. Inf. Forensics Sec. **9**(1), 39–50 (2014)
27. J. Zhou, O.C. Au, G. Zhai, Y.Y. Tang, X. Liu, Scalable compression of stream cipher encrypted images through context-adaptive sampling. IEEE Trans. Inf. Forensics Sec. **9**(11), 1857–1868 (2014)
28. E.J. Candès, J. Romberg, T. Tao, Robust uncertainty principles: exact signal reconstruction from highly incomplete frequency information. IEEE Trans. Inf. Theory **52**(2), 489–509 (2006)
29. D.L. Donoho, Compressed sensing. IEEE Trans. Inf. Theory **52**(4), 1289–1306 (2006)
30. H. Fang, S.A. Vorobyov, H. Jiang, O. Taheri, Permutation meets parallel compressed sensing: how to relax restricted isometry property for 2D sparse signals. IEEE Trans. Signal Process. **62**(1), 196–210 (2014)
31. Y. Zhang, J. Zhou, F. Chen, L.Y. Zhang, K.-W. Wong, H. Xing, D. Xiao, Embedding cryptographic features in compressive sensing. Neurocomput. **205**, 472–480 (2016)
32. L.Y. Zhang, K.-W. Wong, Y. Zhang, J. Zhou, Bi-level protected compressive sampling. IEEE Trans. Multimed. **18**(9), 1720–1732 (2016)
33. V.K. Goyal, A.K. Fletcher, S. Rangan, Compressive sampling and lossy compression. IEEE Signal Process. Mag. **25**(2), 48–56 (2008)
34. M. Grant, S. Boyd, Y. Ye, CVX: Matlab software for disciplined convex programming (2008)
35. J. Katz, Y. Lindell, *Introduction to Modern Cryptography* (CRC press, 2014)

Chapter 6
Concluding Remarks and Future Research

This book investigated the applications of secure CS in multimedia data, cloud computing, and IoT. For multimedia data security, we give some combination frameworks of CS integrated into chaos encryption or optics encryption, designed parallel CS for securing multi-dimensional multimedia data, involved image processing techniques to fulfill scalable encryption framework, and proposed a double protection mechanism against plaintext attacks. The measurement matrix in CS always has a big size and takes up too much space, leading to a lot of inconveniences for communication and sharing. To address this issue, one solution is to generate it using chaotic sequences through iterating certain chaotic system from its initial values. It only requires for transmitting the initial values. Till now, such measurement matrices have been investigated in [1–6], which brings a benefit for secure multimedia CS scheme design like [7]. Not only that, but the chaotic properties such as the sensitivity to initial values, pseudo-randomness and ergodicity can be infused in those CS based multimedia encryption schemes due to being closely connected to cryptographic features. In other words, joint multimedia compression and encryption can be realized by combining CS and chaos, where CS mainly aims at compression and chaos is intended to provide security assurance.

The future research topics on combination CS with optics are twofold. On one hand, in recent years, optical imaging based on CS is a hot topic, which has been emerged in a number of important research results [8–13]. Interestingly, we can attempt the problem of how to embed cryptographic features in these compressive imaging schemes. Secure compressive imaging will further have a wide range of applications. On the other hand, a few joint optical image compression and encryption schemes have been proposed in [14–17]. However, these schemes are only suitable for multiple images. Few references involve the case of a single image, but CS offers such an opportunity. Instead of the simple combination of CS and DRPE [18, 19], the fusion of both will be possibly established to simultaneously compress and encrypt an image, as shown in [20].

© The Author(s), under exclusive license to Springer Nature Singapore Pte Ltd. 2019   113
Y. Zhang et al., *Secure Compressive Sensing in Multimedia Data,*
Cloud Computing and IoT, SpringerBriefs in Signal Processing,
https://doi.org/10.1007/978-981-13-2523-6_6

For cloud computing security, we constructed two secure CS reconstruction protocols in multi-clouds, general sparse reconstruction service and sparse robustness decoding service. For IoT security, we presented a secure low-cost CS scheme a secure data storage and sharing service in IoT. With respect to cloud computing and IoT, they are often inseparable. The storage and management of big data cannot be separated from cloud computing and IoT data are also no exception. Sensor-cloud model [21] is a representative example of both combined with each other. Thus, secure CS sampling in IoT and privacy-preserving CS reconstruction and sharing in cloud should be taken into account simultaneously, which is worthy of study in future. Furthermore, not only in the background of cloud computing, fog computing [22] and edge computing [23] can also be considered and secure CS can be applied due to its advantages including low-cost, robustness, and simultaneous sampling, compression, and encryption.

References

1. L. Yu, J.P. Barbot, G. Zheng, H. Sun, Compressive sensing with chaotic sequence. IEEE Signal Process. Lett. **17**(8), 731–734 (2010)
2. L. Yu, J.-P. Barbot, G. Zheng, H. Sun, Toeplitz-structured chaotic sensing matrix for compressive sensing, in *Proceedings of International Symposium on Communication Systems, Networks and Digital Signal Processing, CSNDSP* (2010), pp. 229–233
3. V. Kafedziski, T. Stojanovski, Compressive sampling with chaotic dynamical systems, in *Proceedings of 19th Telecommunications Forum, TELFOR* (2011), pp. 695–698
4. M. Frunzete, L. Yu, J. Barbot, A. Vlad, Compressive sensing matrix designed by tent map, for secure data transmission, in *Proceedings of IEEE Signal Processing Algorithms, Architectures, Arrangements, and Applications, SPA*, Poznan (2011), pp. 1–6
5. G. Chen, D. Zhang, Q. Chen, D. Zhou, The characteristic of different chaotic sequences for compressive sensing, in *Proceedings of 5th International Congress Image Signal Processing, CISP* (2012), pp. 1475–1479
6. H. Gan, Z. Li, J. Li, X. Wang, Z. Cheng, Compressive sensing using chaotic sequence based on Chebyshev map. Nonlinear Dyn. **78**(4), 2429–2438 (2014)
7. N. Zhou, A. Zhang, F. Zheng, L. Gong, Novel image compression-encryption hybrid algorithm based on key-controlled measurement matrix in compressive sensing. Opt. Laser Techn. **62**, 152–160 (2014)
8. W.-K. Yu, M.-F. Li, X.-R. Yao, X.-F. Liu, L.-A. Wu, G.-J. Zhai, Adaptive compressive ghost imaging based on wavelet trees and sparse representation. Opt. Express **22**(6), 7133–7144 (2014)
9. G. Oliveri, L. Poli, P. Rocca, A. Massa, Bayesian compressive optical imaging within the Rytov approximation. Opt. Lett. **37**(10), 1760–1762 (2012)
10. J. Greenberg, K. Krishnamurthy, D. Brady, Compressive single-pixel snapshot X-ray diffraction imaging. Opt. Lett. **39**(1), 111–114 (2014)
11. X. Lin, G. Wetzstein, Y. Liu, Q. Dai, Dual-coded compressive hyperspectral imaging. Opt. Lett. **39**(7), 2044–2047 (2014)
12. S. Evladov, O. Levi, A. Stern, Progressive compressive imaging from Radon projections. Opt. Express **20**(4), 4260–4271 (2012)
13. H. Shen, L. Gan, N. Newman, Y. Dong, C. Li, Y. Huang, Y. Shen, Spinning disk for compressive imaging. Opt. Lett. **37**(1), 46–48 (2012)
14. A. Alfalou, C. Brosseau, Optical image compression and encryption methods. Adv. Opt. Photonic. **1**(3), 589–636 (2009)

15. A. Alfalou, C. Brosseau, Exploiting root-mean-square time-frequency structure for multiple-image optical compression and encryption. Opt. Lett. **35**(11), 1914–1916 (2010)
16. A. Alfalou, C. Brosseau, N. Abdallah, M. Jridi, Simultaneous fusion, compression, and encryption of multiple images. Opt. Express **19**(24), 24023–24029 (2011)
17. A. Alfalou, A. Loussert, A. Alkholidi, R. El Sawda, System for image compression and encryption by spectrum fusion in order to optimize image transmission, in *Future Generation Communication and Networking*, vol. 2 (IEEE, 2007), pp. 590–593
18. B. Deepan, C. Quan, Y. Wang, C. Tay, Multiple-image encryption by space multiplexing based on compressive sensing and the double-random phase-encoding technique. Appl. Opt. **53**(20), 4539–4547 (2014)
19. X. Liu, W. Mei, H. Du, Optical image encryption based on compressive sensing and chaos in the fractional Fourier domain. J. Modern Opt. **61**(19), 1570–1577 (2014)
20. Y. Zhang, L.Y. Zhang, Exploiting random convolution and random subsampling for image encryption and compression. Electron. Lett. **51**(20), 1572–1574 (2015)
21. A. Alamri, W.S. Ansari, M.M. Hassan, M.S. Hossain, A. Alelaiwi, M.A. Hossain, A survey on sensor-cloud: architecture, applications, and approaches. Int. J. Distri. Sensor Netw. **9**(2), 917923 (2013)
22. F. Bonomi, R. Milito, J. Zhu, S. Addepalli, Fog computing and its role in the internet of things, in *Proceedings of First Edition MCC Workshop Mobile Cloud Computing* (ACM, 2012), pp. 13–16
23. W. Shi, J. Cao, Q. Zhang, Y. Li, L. Xu, Edge computing: vision and challenges. IEEE Internet Things J. **3**(5), 637–646 (2016)

Printed in the United States
By Bookmasters